Current Topics in Microbiology and Immunology

Volume 336

Current Topics in Microbiology and Immunology

Previously published volumes
Further volumes can be found at springer.com

Vol. 311: **Pulendran, Bali; Ahmed, Rafi (Eds.):** From Innate Immunity to Immunological Memory. 2006. 13 figs. X, 177 pp. ISBN 3-540-32635-9

Vol. 312: **Boshoff, Chris; Weiss, Robin A. (Eds.):** Kaposi Sarcoma Herpesvirus: New Perspectives. 2006. 29 figs. XVI, 330 pp. ISBN 3-540-34343-1

Vol. 313: **Pandolfi, Pier P.; Vogt, Peter K.(Eds.):** Acute Promyelocytic Leukemia. 2007. 16 figs. VIII, 273 pp. ISBN 3-540-34592-2

Vol. 314: **Moody, Branch D. (Ed.):** T Cell Activation by CD1 and Lipid Antigens, 2007, 25 figs. VIII, 348 pp. ISBN 978-3-540-69510-3

Vol. 315: **Childs, James, E.; Mackenzie, John S.; Richt, Jürgen A. (Eds.):** Wildlife and Emerging Zoonotic Diseases: The Biology, Circumstances and Consequences of Cross-Species Transmission. 2007. 49 figs. VII, 524 pp. ISBN 978-3-540-70961-9

Vol. 316: **Pitha, Paula M. (Ed.):** Interferon: The 50th Anniversary. 2007. VII, 391 pp. ISBN 978-3-540-71328-9

Vol. 317: **Dessain, Scott K. (Ed.):** Human Antibody Therapeutics for Viral Disease. 2007. XI, 202 pp. ISBN 978-3-540-72144-4

Vol. 318: **Rodriguez, Moses (Ed.):** Advances in Multiple Sclerosis and Experimental Demyelinating Diseases. 2008. XIV, 376 pp. ISBN 978-3-540-73679-9

Vol. 319: **Manser, Tim (Ed.):** Specialization and Complementation of Humoral Immune Responses to Infection. 2008. XII, 174 pp. ISBN 978-3-540-73899-2

Vol. 320: **Paddison, Patrick J.; Vogt, Peter K.(Eds.):** RNA Interference. 2008. VIII, 273 pp. ISBN 978-3-540-75156-4

Vol. 321: **Beutler, Bruce (Ed.):** Immunology, Phenotype First: How Mutations Have Established New Principles and Pathways in Immunology. 2008. XIV, 221 pp. ISBN 978-3-540-75202-8

Vol. 322: **Romeo, Tony (Ed.):** Bacterial Biofilms. 2008. XII, 299. ISBN 978-3-540-75417-6

Vol. 323: **Tracy, Steven; Oberste, M. Steven; Drescher, Kristen M. (Eds.):** Group B Coxsackieviruses. 2008. ISBN 978-3-540-75545-6

Vol. 324: **Nomura, Tatsuji; Watanabe, Takeshi; Habu, Sonoko (Eds.):** Humanized Mice. 2008. ISBN 978-3-540-75646-0

Vol. 325: **Shenk, Thomas E.; Stinski, Mark F. (Eds.):** Human Cytomegalovirus. 2008. ISBN 978-3-540-77348-1

Vol. 326: **Reddy, Anireddy S.N; Golovkin, Maxim (Eds.):** Nuclear pre-mRNA processing in plants. 2008. ISBN 978-3-540-76775-6

Vol. 327: **Manchester, Marianne; Steinmetz, Nicole F. (Eds.):** Viruses and Nanotechnology. 2008. ISBN 978-3-540-69376-5

Vol. 328: **van Etten, (Ed.):** Lesser Known Large dsDNA Viruses. 2008. ISBN 978-3-540-68617-0

Vol. 329: **Diane E. Griffin; Michael B.A. Oldstone (Eds.):** Measles 2009. ISBN 978-3-540-70522-2

Vol. 330: **Diane E. Griffin; Michael B.A. Oldstone (Eds.):** Measles 2009. ISBN 978-3-540-70616-8

Vol. 331: **Villiers, E. M. de (Eds):** TT Viruses. 2009. ISBN 978-3-540-70917-8

Vol. 332: **Karasev A.(Ed.):** Plant produced Microbial Vaccines. 2009. ISBN 978-3-540- 70857-5

Vol. 333: **Richard W. Compans; Walter A. Orenstein (Eds):** Vaccines for Pandemic Influenza. 2009. ISBN 978-3-540-92164-6

Vol. 334: **Dorian McGavern; Micheal Dustin (Eds.):** visualizing Immunity. 2009. ISBN 978-3-540-93862-0

Vol. 335: **Beth Levine; Tamotsu Yoshimori; Vojo Deretic (Eds.):** Autophagy in Infection and Immunity. 2009. ISBN 978-3-642-00301-1

Tammy Kielian
Editor

Toll-like Receptors: Roles in Infection and Neuropathology

Springer

Editor
Tammy Kielian
Department of Pathology
and Microbiology
University of Nebraska
Medical Center
NE, USA
tkielian@nmc.edu

ISBN 978-3-642-00548-0 e-ISBN 978-3-642-00549-7
DOI 10.1007/978-3-642-00549-7
Springer Dordrecht Heidelberg London New York

Current Topics in Microbiology and Immunology ISSN 0070-217x

Library of Congress Catalog Number: 2009926942

Cover design: WMX Design GmbH, Heidelberg, Germany

Printed on acid-free paper

Springer is part of Springer Science+Business Media (www.springer.com)

This book is dedicated to B.B., who was at my side during the writing of most of my scientific works.

Contents

Contributors

Celia F. Brosnan
Department of Pathology (Neuropathology), Albert Einstein College of Medicine, 1300 Morris Park Avenue, Bronx, NY 10461, USA

Paul D. Drew
Department of Neurobiology and Developmental Sciences, University of Arkansas for Medical Sciences, Little Rock, AR 72205, USA
DrewPaulD@uams.edu

Nilufer Esen
Department of Neurology, University of Michigan Medical School, Ann Arbor, MI, USA

Uma Mahesh Gundra
Department of Biology, South Texas Center for Emerging Infectious Diseases, The University of Texas at San Antonio, One UTSA Circle, San Antonio, TX 78249-1644, USA

Tammy Kielian
Department of Pathology and Microbiology, University of Nebraska Medical Center, Omaha, NE, USA
tkielian@unmc.edu

Kristina A. Kigerl
Center for Brain and Spinal Cord Repair, Department of Molecular Virology, Immunology and Medical Genetics, The Ohio State University College of Medicine, Columbus, OH, USA

Donghoon Kim
Program in Molecular and Cellular Neuroscience, DRI, BK21, and Department of Oral Physiology, School of Dentistry, Seoul National University, Seoul, Republic of Korea

Uwe Koedel
Department of Neurology, Clinic of the Ludwig-Maximilians, University of Munich, Marchioninistr 15, Munich 81377, Germany
Uwe.Koedel@med.uni-muenchen.de

Gary E. Landreth
Alzheimer Research Laboratory, Department of Neurosciences, Case Western Reserve University, Cleveland, OH 44106, USA
gel2@case.edu

Sunhee C. Lee
Department of Pathology (Neuropathology), Albert Einstein College of Medicine, 1300 Morris Park Avenue, Bronx, NY 10461, USA
slee@aecom.yu.edu

Soojin Lee
Department of Microbiology, Chungnam National University, Daejoen, Republic of Korea
sjlee87@snu.ac.kr

Sung Joong Lee
Program in Molecular and Cellular Neuroscience, DRI, BK21, and Department of Oral Physiology, School of Dentistry, Seoul National University, Seoul, Republic of Korea

Bibhuti B. Mishra
Department of Biology, South Texas Center for Emerging Infectious Diseases, The University of Texas at San Antonio, One UTSA Circle, San Antonio, TX 78249-1644, USA

Trevor Owens
Medical Biotechnology Center, University of Southern Denmark, Winsloewparken 25, 5000, Odense C, Denmark
towens@health.sdu.dk

Phillip G. Popovich
Department of Molecular Virology, Immunology and Medical Genetics, Center for Brain and Spinal Cord Repair, The Ohio State University College of Medicine, Columbus, OH, USA
Phillip.Popovich@osumc.edu

Michael K. Racke
Department of Neurology and Department of Molecular Virology, Immunology and Medical Genetics, The Ohio State University Medical Center, 1654 Upham Drive, 445 Means Hall, Columbus, OH 43210, USA

Erin G. Reed-Geaghan
Alzheimer Research Laboratory, Department of Neurosciences, Case Western Reserve University, Cleveland, OH 44106, USA
egr3@case.edu

Hyeon-Sook Suh
Department of Pathology (Neuropathology), Albert Einstein College of Medicine, 1300 Morris Park Avenue, Bronx, NY 10461, USA

Judy M. Teale
Department of Biology, South Texas Center for Emerging Infectious Diseases,
The University of Texas at San Antonio, One UTSA Circle, San Antonio,
TX 78249-1644, USA
Judy.Teale@utsa.edu

Overview of Toll-Like Receptors in the CNS

Tammy Kielian

Contents

Abstract Mammalian Toll-like receptors (TLRs) were first identified in 1997 based on their homology with *Drosophila Toll*, which mediates innate immunity in the fly. Over the past eight years, the number of manuscripts describing TLR expression and function in the central nervous system (CNS) has been increasing steadily and expanding beyond their traditional roles in infectious diseases to neurodegenerative disorders and injury. Interest in the field serves as the impetus for this volume in the *Current Topics in Microbiology and Immunology* series entitled *Toll-Like Receptors: Roles in Infection and Neuropathology*. The first five chapters highlight more traditional roles for TLRs in infectious diseases of the CNS. The second half of the volume discusses recently emerging roles for TLRs in noninfectious neurodegenerative diseases and the challenges faced by these models in identifying endogenous ligands. Several conceptual theories are introduced in various chapters that deal with the dual nature of TLR engagement and whether these signals favor neuroprotective versus neurodegenerative outcomes.

T. Kielian
Department of Pathology and Microbiology, University of Nebraska Medical Center, Omaha, NE, USA
e-mail: tkielian@unmc.edu

T. Kielian (ed.), *Toll-like Receptors: Roles in Infection and Neuropathology*,
Current Topics in Microbiology and Immunology 336,
DOI 10.1007/978-3-642-00549-7_1,

Abbreviations

APC	Antigen-presenting cell
CNS	Central nervous system
CTMI	*Current Topics in Microbiology and Immunology*
DAMP	Danger-associated molecular pattern
dsRNA	Double-stranded RNA
FACS	Fluorescent-activated cell sorting
IFN	Interferon
IκB	Inhibitory kappa B
IκK	Inhibitory-κB kinase
IL	Interleukin
IL-1R	Interleukin-1 receptor
IL-18R	Interleukin-18 receptor
IRAK	Interleukin-1 receptor-associated kinase
LPS	Lipopolysaccharide
LTA	Lipoteichoic acid
MAPK	Mitogen-activated protein kinase
MyD88	Myeloid differentiation primary-response protein 88
NF-κB	Nuclear factor kappa B
NIK	NF-κB inducing kinase
ODN	Oligodeoxynucleotide
Pam3Cys	Tripalmitoyl-*S*-glyceryl-cysteine
PAMP	Pathogen-associated molecular pattern
PGN	Peptidoglycan
Poly I:C	Polyinosine:cytosine
PRR	Pattern recognition receptor
ssRNA	Single-stranded RNA
TIRAP	Toll-interleukin 1 receptor (TIR) domain-containing adaptor protein
TLR	Toll-like receptor
TRAF	Tumor necrosis factor receptor-associated factor
TRIF	TIR-domain-containing adaptor inducing interferon-β

1 Historical Background of TLRs

The *Toll* gene was first identified in *Drosophila* when a mutation introduced into the gene led to defects in dorsal-ventral patterning in the fly and an inability to completely coalesce the abdominal cavity (Lemaitre et al. 1996). Fortuitously, the discovery was also made that the *Toll* mutation resulted in enhanced susceptibility to fungal infections, providing the first clue that this receptor may participate in the fly innate immune response. Additional evidence came with the finding that a mutation in a distinct *Drosophila* receptor related to Toll, 18-wheeler, led to an increased prevalence of bacterial infections (Williams et al. 1997), reinforcing the pivotal roles that these receptors play in fly immunity. Subsequent sequencing of both the *Toll* and *18-wheeler* genes revealed a significant degree of homology in the cytoplasmic tail with the cytoplasmic domain of the mammalian IL-1R (i.e., >90%) (Anderson 2000). Based on this extraordinary degree of similarity, the search for mammalian homologs of Toll began, and in 1997 the laboratory of the late Dr. Charles Janeway was the first to discover a Toll homolog in human monocytes, namely TLR4 (Medzhitov et al. 1997). Subsequent studies revealed that TLR4 was the receptor for LPS, a major immunostimulatory component of the outer cell wall of Gram-negative bacteria that had remained elusive for over 40 years. Later reports demonstrated that natural mutations in TLR4 were responsible for the LPS hyporesponsive nature of C3H/HeJ and B10 mice (Poltorak et al. 1998a; Hoshino et al. 1999; Qureshi et al. 1999). To date, a total of 13 TLRs have been identified in mice and ten in humans, although ligands for a few of these receptors remain to be defined.

Innate immunity represents the first line of defense against invading microbes. In contrast, the adaptive immune response directed against microbial antigens takes several days to become established. Unlike adaptive immunity, where T and B cells can recognize an infinite repertoire of antigens due to random gene rearrangements of their receptors, cells of the innate immune system rely on a restricted set of germline-encoded receptors that are directed against highly conserved motifs expressed by large classes of microorganisms. These conserved motifs have been coined pathogen-associated molecular patterns (PAMPs), and the receptors that recognize these structures are referred to as pattern recognition receptors (PRRs) (Medzhitov and Janeway 2000; Qureshi and Medzhitov 2003; Kaisho and Akira 2004). In addition to traditional PAMPs, recent evidence has indicated that TLRs can recognize an array of endogenous molecules that are typically sequestered from the immune response, so called "danger signals" or danger-associated molecular patterns (DAMPs) (Matzinger 2002). In this issue of *Current Topics in Microbiology and Immunology*, chapters dedicated to both pathogen-derived and endogenous TLR ligands will be discussed. However, it is likely that during the course of CNS infectious diseases, self-antigens will be liberated as a consequence of cell death/necrosis, and as such, TLR engagement may be elicited by a combination of PAMPs and DAMPs in this context.

In addition to driving innate immune responses to infectious pathogens, TLR-dependent signaling also initiates adaptive immunity (Hoebe et al. 2004; Pasare and Medzhitov 2004). This is particularly evident when considering that the engagement of TLRs by bacterial antigens is required to induce co-stimulatory molecule expression on antigen-presenting cells (APC; i.e., dendritic cells and macrophages) for subsequent activation and expansion of antigen-specific T cells (Hertz et al. 2001; Boonstra et al. 2003; Pasare and Medzhitov 2003, 2004; Hoebe et al. 2004). In addition, cytokines released by TLR-activated APCs, such as IL-12, play a pivotal role in regulating T cell development (Hoebe et al. 2004).

2 TLR Subtypes and Ligand Classification

Currently, there are 13 TLR family members that have been described in mice and 11 in humans, although some still remain relatively poorly characterized (i.e., TLR11, TLR12, TLR13). In terms of this introductory chapter, a discussion of classical microbial TLR ligands will be presented, followed by a description of endogenous TLR agonists that have been more recently described. This presentation order follows the organization of chapters in this volume: the first portion of the book deals with the roles of TLRs in CNS infectious diseases, whereas the second portion addresses new emerging evidence that TLR signaling impacts the course of neurodegenerative disorders where pathogen etiologies are not apparent.

2.1 Extracellular TLRs and CNS Expression Patterns

There are several extracellular TLRs that recognize conserved structural motifs of large microbe populations. These motifs are typically less likely to undergo mutation since they are essential for pathogen survival. The most well-characterized and studied extracellular TLRs are TLR2 and TLR4, which recognize bacterial peptidoglycan (PGN)/lipoproteins and LPS, respectively. For the purposes of this introductory chapter, only these receptors will be discussed; however, the reader is directed to several excellent review articles on the subject for more information (Akira et al. 2006; O'Neill and Bowie 2007).

TLR2 is capable of recognizing the widest array of PAMPs identified to date, including PGN, bacterial lipoproteins (i.e., tripalmitoyl-*S*-glyceryl-cysteine; Pam3Cys), atypical LPS from *Prophyromonas gingivitis* and *Leptospira interrogans*, glycosylphosphotidylinositol lipid from *Trypanosoma cruzi*, and yeast zymosan (Qureshi and Medzhitov 2003; Takeda et al. 2003).

Several studies have demonstrated that microglia express TLR2 (Laflamme et al. 2001, 2003; Bsibsi et al. 2002; Kielian et al. 2002, 2005; Rasley et al. 2002; Zekki et al. 2002; Olson and Miller 2004) and receptor expression is elevated in response to a wide array of TLR2 agonists, including PGN and Pam3Cys as well as alternative TLR ligands (i.e., LPS) (Laflamme et al. 2001, 2003; Rasley et al. 2002; Olson and Miller 2004). Another PRR that has been reported to cooperate with TLR2 is CD14, which is expressed on cells of the myeloid lineage including microglia and macrophages (Becher et al. 1996; Nadeau and Rivest 2000; Saito et al. 2000; Kielian et al. 2002, 2005). CD14 is a glycosylphosphatidyl inositol (GPI)-anchored receptor and is involved in the recognition of Gram-positive PAMPs such as PGN and LTA through its ability to interact with TLR2/TLR1 and/or TLR2/TLR6 heterodimers (Cleveland et al. 1996; Gupta et al. 1996; Dziarski et al. 2000; Henneke et al. 2001; Schroder et al. 2003; Weber et al. 2003; Manukyan et al. 2005). More classically, CD14 is known for its ability to pair with TLR4 to transduce activation signals in response to LPS (Haziot et al. 1988; Dobrovolskaia and Vogel 2002; Fitzgerald et al. 2004; Palsson-McDermott and O'Neill 2004).

Astrocytes also express TLR2, with augmented receptor levels observed upon exposure to various PAMPs (Bowman et al. 2003; Esen et al. 2004; Carpentier et al. 2005). Strong evidence demonstrating that astrocytes express TLR2 in vivo was shown by Mishra et al. using immunofluorescence staining. In this study, robust TLR2 immunoreactivity was detected in astrocytes in both the normal and infected CNS (Mishra et al. 2006). Further support for astrocytic TLR2 expression was provided by a recent report by Kigerl et al. that utilized laser capture microdissection for astrocyte enrichment from control and injured spinal cord tissues and demonstrated TLR2 expression associated with astrocytes, although maximal expression was detected in microglia (Kigerl et al. 2007). However, studies examining TLR2 in other systems have produced some conflicting results with regard to astrocytic expression (Bsibsi et al. 2002; Rivest 2003; Farina et al. 2005; Owens 2005). It is likely that the context of PAMP exposure and/or the strength of the activation signal received following astrocyte activation may dictate whether TLR2 expression is induced. This "strength of signal" concept is proposed by Trevor Owens in the chapter "Toll-Like Receptors in Neurodegeneration" to address these discrepancies. Alternative explanations may include the species from which astrocytes were procured, the route of PAMP administration during in vivo studies, and/or the length of time that astrocytes are co-cultured with microglia prior to purification for in vitro studies.

As mentioned earlier, TLR4 is responsible for recognizing the Gram-negative cell wall component LPS (Poltorak et al. 1998a,b; Heine et al. 1999; Hoshino et al. 1999; Qureshi et al. 1999; Takeuchi et al. 1999; Hirschfeld et al. 2000; Lien et al. 2000; Tapping et al. 2000). With respect to glia, it has long been acknowledged that LPS serves as a potent stimulus for microglial activation typified by the robust production of numerous proinflammatory mediators. Therefore, it was not unexpected when microglia were reported to express TLR4 (Laflamme and Rivest 2001; Bsibsi et al. 2002; Lehnardt et al. 2002, 2003; Laflamme et al. 2003; Rivest 2003; Olson and Miller 2004; Chakravarty and Herkenham 2005; Jung et al. 2005). As previously mentioned, CD14 interacts with TLR4 to induce maximal responses to LPS in macrophages and microglia (Dobrovolskaia and Vogel 2002; O'Neill 2004; Palsson-McDermott and O'Neill 2004; Esen and Kielian 2005).

In contrast to microglia, it appears more controversial as to whether astrocytes express TLR4. Several groups have been unable to demonstrate astrocytic TLR4 expression in vitro (Farina et al. 2005; Kielian, unpublished observations) or in vivo (Laflamme and Rivest 2001; Lehnardt et al. 2002, 2003); however, others have detected low, constitutive expression of TLR4 in astrocytes that is increased upon cell activation (Bsibsi et al. 2002; Bowman et al. 2003; Carpentier et al. 2005). It is important to acknowledge that great care must be taken when working with primary astrocytes to ensure that contamination with microglia is relatively low (Saura 2007). Since microglia express high levels of TLR4, a small number of residual microglia could introduce artifact signals that are not reflective of astrocytic receptor expression. This topic is also discussed in the chapter "Toll-Like

Receptors in Neurodegeneration" in this volume. Further studies using primary astrocyte cultures where microglia have been depleted by immunological means (i.e., magnetic bead purification or FACS using CD11b) will help to resolve this lingering issue.

2.2 *Intracellular TLRs and CNS Expression Patterns*

Not all TLRs are expressed at the plasma membrane; several—including TLR3, TLR7/8, and TLR9—are associated with endosomal membranes intracellularly. The intracellular expression patterns of these TLRs appear logical given the fact that their ligands represent nucleic acid motifs of pathogens that are typically not found extracellularly. Indeed, these nucleic acid motifs are typically encountered during intracellular replication and/or within intracellular compartments following phagocytosis. TLR3 recognizes dsRNA, which is an intermediate produced during viral replication in cells (Alexopoulou et al. 2001). Studies investigating the potential role of TLR3-mediated signaling commonly utilize the synthetic TLR3 agonist polyinosine:cytosine (poly I:C); however, TLR3 expression does not appear to be regulated by poly I:C in microglia (Bsibsi et al. 2002; Olson and Miller 2004), which differs from some of the other TLRs where receptor levels are augmented following exposure to their natural agonist(s) (Olson and Miller 2004; Kielian et al. 2005). Unlike the discrepancy in TLR4 expression, there is a consensus that astrocytes do express TLR3 (Bsibsi et al. 2002; Carpentier et al. 2005; Farina et al. 2005, 2007; Scumpia et al. 2005). A central role for astrocytes in sensing viral infections in the CNS is supported by the finding that cells are responsive to the TLR3 agonist poly I:C, as is made evident by the production of several proinflammatory mediators.

TLR7 and TLR8 are highly homologous TLRs and their ligands include single-stranded RNA (ssRNA) as well as structurally similar synthetic chemicals including antiviral and anticancer compounds (Kaisho and Akira 2004). TLR7 and TLR8 expression has been reported in microglia (Bsibsi et al. 2002; Olson and Miller 2004), astrocytes (Carpentier et al. 2005), and more recently neurons (Ma et al. 2006, 2007), where TLR8 expression drives neuronal phenotypic changes and regulates apoptosis.

TLR9 mediates responses to bacterial DNA, viral DNA, and synthetic oligodeoxynucleotides (ODN), all of which contain unmethylated CpG motifs (Takeda et al. 2003). Both microglia (Takeshita et al. 2001; Dalpke et al. 2002; Iliev et al. 2004; Olson and Miller 2004; Zhang et al. 2005) and astrocytes (Bowman et al. 2003; Hosoi et al. 2004; Carpentier et al. 2005) express TLR9, and engagement of this PRR leads to a robust induction of proinflammatory mediators.

Following pathogen infection, it is likely that these intracellular TLRs serve to amplify the host immune response that was initially triggered by extracellular TLRs to ensure effective pathogen clearance. However, in the context of noninfectious neurodegeneration, the pathologic engagement of intracellular

TLRs by endogenous ligands may contribute to exacerbated immune responses and enhance neuropathology. These issues are discussed in the chapters "Toll-Like Receptors in Neurodegeneration," "Toll-Like Receptors in Spinal Cord Injury," " Toll-Like Receptors in Alzheimer's Disease," "Toll-Like Receptors in Multiple Sclerosis," and "Toll-Like Receptors in Peripheral Nerve Injury and Neuropathic Pain" in this *CMTI* volume.

Recently, several studies have described endogenous molecules that are capable of triggering TLR-dependent signaling cascades (Tsan and Gao 2004). One issue that has confounded progress in this area is the concern of reagent purity; in particular, earlier studies describing TLR-dependent signaling pathways for endogenous molecules were complicated by contaminating LPS (Tsan and Gao 2004). However, in spite of this issue, convincing evidence has emerged documenting the ability of several endogenous molecules to engage TLRs, with the majority stimulating either TLR2 or TLR4 (Tsan and Gao 2004; Kielian 2006). Despite the fact that several models of CNS injury have been shown to be influenced by TLR2 and/or TLR4-dependent signaling, the identity of the ligand(s) that trigger these receptors remains elusive.

3 TLR Signaling Pathways

TLR engagement culminates in the induction of NF-κB and MAPK signaling pathways, both of which regulate the expression of a wide array of genes involved in immune responses. Since the majority of TLRs utilize the central adaptor molecule MyD88 to transduce signaling cascades, this scheme will be discussed briefly with differences in TLR3-dependent signaling to follow (Akira 2006; O'Neill and Bowie 2007). TLR activation results in the recruitment of the adaptor protein MyD88, which is associated with the serine/threonine kinase interleukin-1 receptor-associated kinase (IRAK). Subsequently, IRAK interacts with TNF receptor-associated factor (TRAF) adaptor protein TRAF6, which provides a bridge to the protein kinase NF-κB-inducing kinase (NIK). Next, NIK phosphorylates IKK (IκB kinase), leading to IκB phosphorylation. IκB phosphorylation targets the protein for ubiquitination and proteasome-mediated degradation, resulting in the release and nuclear translocation of NF-κB, whereupon it can influence the expression of numerous immune response genes. However, recent evidence has demonstrated the existence of alternative adaptor molecules that transduce signals from TLRs via a MyD88-independent pathway (Akira and Takeda 2004). These adaptors include TRAM and TRIF, which are pivotal for the expression of IFN-inducible genes following TLR4 activation (Yamamoto et al. 2003, 2004; Akira and Takeda 2004). TRIF is also required for TLR3-mediated signaling in response to dsRNA and is responsible for the induction of type I interferons (i.e., IFN-α and IFN-β) that are a hallmark host innate immune response to viral infection.

4 Highlights of Contributing Chapters and Emerging Concepts

The objective of this volume is to provide a current synopsis on the role of TLRs during both infectious and noninfectious diseases affecting the CNS. Traditionally, TLRs have been regarded as pathogen sensors and, as such, the early TLR literature in the CNS was focused on this topic. However, recent studies utilizing various TLR-deficient mouse strains have revealed that TLRs can also impact the course of distinct neurodegenerative diseases/pathologies. Although the ligands responsible for triggering TLR involvement in the absence of infectious insults have not yet been elucidated, it is apparent that these PRRs play a role, at some level, in influencing the subsequent host immune response to injury/trauma and the subsequent regenerative response. It is anticipated that this book will serve as a forum to bring to light the various outstanding questions that remain in the field as well as to introduce new concepts regarding the roles of TLRs in CNS diseases and, importantly, acknowledge the complexity of TLR signaling and the likelihood that TLRs act in concert with additional receptors to orchestrate the subsequent inflammatory profile.

The first four chapters of this book address the roles of TLRs in various models of CNS infectious disease, including bacterial meningitis, brain abscess, and viral and parasitic infections that target the brain. Several interesting concepts emerge from these discussions that emphasize the relatedness of TLR involvement despite the distinct infectious etiologies and diverse TLR engagement employed. This suggests that infectious insults may elicit a "common initial pathway" for inflammation that can be further refined to accomplish the outcome required to neutralize the specific pathogen. This would translate to an early conserved innate immune response followed by a tailored pathogen-specific cascade. This concept remains to be supported or refuted, but nonetheless, comparisons between diverse infectious disease models should be made and the results utilized to make such determinations. This is one objective of assimilating these chapters into one volume.

Another commonality shared between the various infectious disease models presented in this book is the fact that the resultant immune response (mediated, in part, via TLRs) not only leads to pathogen destruction but also bystander damage to surrounding CNS parenchyma by necrotic/apoptotic cell death. Therefore, it is also possible that during CNS infections, TLRs play two roles in ligand recognition: 1) to facilitate the initial response to the inciting pathogen and 2) upon tissue destruction, TLRs may also recognize newly liberated self-antigens as a result of necrotic cell death, a so-called "pathogen-necrosis-autoantigen triad" that is proposed in the chapter "Toll-Like Receptors in Brain Abscess." This could conceivably account for the exaggerated inflammatory response that typically accompanies these CNS infectious disorders. This concept remains to be tested; however, it remains an intriguing area for future investigation. It is clear from studies described in subsequent chapters of this book that agonist(s) liberated following CNS injury/insult are indeed capable of interfacing with TLRs to modulate the host response to damage. It remains to be seen whether a protective anti-pathogen response could be dissociated from a potential deleterious anti-self response following the liberation of endogenous TLR ligands to minimize damage to surrounding normal CNS tissue during these infectious insults.

When comparing the roles of TLRs in bacterial meningitis (chapter "Toll-Like Receptors in Bacterial Meningitis") versus brain abscess (chapter "Toll-Like Receptors in Brain Abscess"), although both infections occur in a distinct CNS compartment and involve different responding effector cells during the initial phase of disease, remarkably there are some commonalities observed. For example, both TLR2- and MyD88-dependent signals influence the pathogenesis of both infections. There is more consistency between the two models in terms of MyD88, where the loss of signaling leads to dramatic defects in host innate immunity and the failure to clear infection. In contrast, although TLR2-deficicient mice do exhibit some deficits in bacterial clearance and depressed proinflammatory mediator production, these mice do not experience overt clinical decline as compared to MyD88-deficient animals in either the bacterial meningitis or brain abscess models. These findings implicate the involvement of additional PRRs that participate in bacterial recognition and amplification of immune networks. In addition, the cross-talk between TLRs and phagocytic PRRs must also be addressed, since changes in cytokine expression patterns can influence phagocytic indices (Mukhopadhyay et al. 2004; Underhill and Gantner 2004). Future studies utilizing mice that are deficient for both a particular TLR and phagocytic PRR would be interesting to test the interplay between these molecules.

Another example of the intriguing complexity that could impact TLR signaling during CNS infections is illustrated by helminth diseases, which is the topic of the chapter "Toll-Like Receptors in CNS Parasitic Infections" in this volume. Namely, many parasites that target the CNS harbor their own commensal bacteria. It is intriguing to speculate that parasite death results in a complex milieu of TLR ligands that not only originate from the parasite itself but also its endogenous microflora.

One intriguing observation that has emerged from studies of various models of neurodegeneration/injury is the finding of microglia/macrophage heterogeneity that is dependent on the context of inflammation. This concept is illuminated in the chapter "Toll-Like Receptors in Spinal Cord Injury" with regard to the divergent ability of infiltrating macrophages following traumatic spinal cord injury to exhibit either neurodestructive or neuroprotective properties. It has been proposed by the authors that TLR2 and TLR4 may favor inflammation during the acute stage of injury; however, these same TLRs may serve to promote repair processes/recovery during the later phases of disease (chapter "Toll-Like Receptors in Spinal Cord Injury"). Another explanation is provided in the chapter "Toll-Like Receptors in Neurodegeneration," where a "strength of signal" hypothesis is introduced. This concept states that neuroprotective versus detrimental effects of microglia in the CNS are dictated by the concentration of TLR agonists that these cells are exposed to in vivo (chapter "Toll-Like Receptors in Neurodegeneration"). It is important to acknowledge that both lines of thought are not mutually exclusive and raise important concepts that warrant further investigation in experimental systems.

Additional complexities regarding TLR signaling are raised in the chapter "Toll-Like Receptors in Alzheimer's Disease". In particular, it is evident that extensive receptor complexity exists involving TLRs and CD14 in the recognition of β-amyloid and dictating whether phagocytosis versus proinflammatory mediator production is induced. Likewise, the chapter "Toll-Like Receptors in Multiple Sclerosis" reveals divergent roles for MyD88-dependent versus -independent signaling in either exacerbating or attenuating disease severity in rodent models of multiple sclerosis (i.e., experimental autoimmune encephalomyelitis), respectively.

The chapter "Toll-Like Receptors in Peripheral Nerve Injury and Neuropathic Pain" highlights recent evidence implicating TLR signaling in mediating nerve degeneration/regeneration and neuropathic pain following nerve injury. In common with the other contributions covering topics of neurodegeneration/injury, it remains to be determined which factor(s) dictates whether TLR signaling is beneficial for reparative processes or rather induces pathology and chronic pain responses. Teasing apart these mechanisms may afford new therapeutic treatment modalities for the management of neuropathic pain, which represents a significant socioeconomic burden.

Finally it is important to remind the reader that although it is tempting to assign the phenotypes obtained with MyD88-deficient mice to TLRs, this conclusion cannot be assumed. This is namely because MyD88 is utilized for signaling via the IL-1R and IL-18R as well as TLRs, confounding the interpretations that can be made. This is all the more relevant since studies in models of CNS infectious diseases have demonstrated important roles for IL-1 and IL-18 in the host antibacterial immune response (Zwijnenburg et al. 2003a,b; Kielian et al. 2004). Future studies using mice that are deficient for two TLRs that are suspected to influence the course of disease are needed, or alternatively, animals could be engineered that lack both the IL-1R/or IL-18R and a particular TLR of interest.

Acknowledgments Dr. Kielian's laboratory is supported by grants from the NIH National Institute of Neurological Disorders and Stroke (RO1s NS055385, NS40730, and NS053487).

References

Akira S (2006) TLR signaling. Curr Top Microbiol Immunol 311:1–16

Akira S, Takeda K (2004) Toll-like receptor signalling. Nat Rev Immunol 4:499–511

Akira S, Uematsu S, Takeuchi O (2006) Pathogen recognition and innate immunity. Cell 124:783–801

Alexopoulou L, Holt AC, Medzhitov R, Flavell RA (2001) Recognition of double-stranded RNA and activation of NF-kappaB by Toll-like receptor 3. Nature 413:732–738

Anderson KV (2000) Toll signaling pathways in the innate immune response. Curr Opin Immunol 12:13–19

Becher B, Fedorowicz V, Antel JP (1996) Regulation of CD14 expression on human adult central nervous system-derived microglia. J Neurosci Res 45:375–381

Boonstra A, Asselin-Paturel C, Gilliet M, Crain C, Trinchieri G, Liu YJ, O'Garra A (2003) Flexibility of mouse classical and plasmacytoid-derived dendritic cells in directing T helper

type 1 and 2 cell development: dependency on antigen dose and differential toll-like receptor ligation. J Exp Med 197:101–109
Bowman CC, Rasley A, Tranguch SL, Marriott I (2003) Cultured astrocytes express toll-like receptors for bacterial products. Glia 43:281–291
Bsibsi M, Ravid R, Gveric D, van Noort JM (2002) Broad expression of Toll-like receptors in the human central nervous system. J Neuropathol Exp Neurol 61(11):1013–1021
Carpentier PA, Begolka WS, Olson JK, Elhofy A, Karpus WJ, Miller SD (2005) Differential activation of astrocytes by innate and adaptive immune stimuli. Glia 49:360–374
Chakravarty S, Herkenham M (2005) Toll-like receptor 4 on nonhematopoietic cells sustains CNS inflammation during endotoxemia, independent of systemic cytokines. J Neurosci 25:1788–1796
Cleveland MG, Gorham JD, Murphy TL, Tuomanen E, Murphy KM (1996) Lipoteichoic acid preparations of gram-positive bacteria induce interleukin-12 through a CD14-dependent pathway. Infect Immun 64:1906–1912
Dalpke AH, Schafer MK, Frey M, Zimmermann S, Tebbe J, Weihe E, Heeg K (2002) Immunostimulatory CpG-DNA activates murine microglia. J Immunol 168:4854–4863
Dobrovolskaia MA, Vogel SN (2002) Toll receptors, CD14, and macrophage activation and deactivation by LPS. Microbes Infect 4:903–914
Dziarski R, Ulmer AJ, Gupta D (2000) Interactions of CD14 with components of gram-positive bacteria. Chem Immunol 74:83–107
Esen N, Kielian T (2005) Recognition of *Staphylococcus aureus*-derived peptidoglycan (PGN) but not intact bacteria is mediated by CD14 in microglia. J Neuroimmunol 170:93–104
Esen N, Tanga FY, DeLeo JA, Kielian T (2004) Toll-like receptor 2 (TLR2) mediates astrocyte activation in response to the Gram-positive bacterium *Staphylococcus aureus*. J Neurochem 88:746–758
Farina C, Aloisi F, Meinl E (2007) Astrocytes are active players in cerebral innate immunity. Trends Immunol 28:138–145
Farina C, Krumbholz M, Giese T, Hartmann G, Aloisi F, Meinl E (2005) Preferential expression and function of Toll-like receptor 3 in human astrocytes. J Neuroimmunol 159:12–19
Fitzgerald KA, Rowe DC, Golenbock DT (2004) Endotoxin recognition and signal transduction by the TLR4/MD2-complex. Microbes Infect 6:1361–1367
Gupta D, Kirkland TN, Viriyakosol S, Dziarski R (1996) CD14 is a cell-activating receptor for bacterial peptidoglycan. J Biol Chem 271:23310–23316
Haziot A, Chen S, Ferrero E, Low MG, Silber R, Goyert SM (1988) The monocyte differentiation antigen, CD14, is anchored to the cell membrane by a phosphatidylinositol linkage. J Immunol 141:547–552
Heine H, Kirschning CJ, Lien E, Monks BG, Rothe M, Golenbock DT (1999) Cutting edge:cells that carry A null allele for toll-like receptor 2 are capable of responding to endotoxin. J Immunol 162:6971–6975
Henneke P, Takeuchi O, van Strijp JA, Guttormsen HK, Smith JA, Schromm AB, Espevik TA, Akira S, Nizet V, Kasper DL, Golenbock DT (2001) Novel engagement of CD14 and multiple toll-like receptors by group B streptococci. J Immunol 167:7069–7076
Hertz CJ, Kiertscher SM, Godowski PJ, Bouis DA, Norgard MV, Roth MD, Modlin RL (2001) Microbial lipopeptides stimulate dendritic cell maturation via Toll-like receptor 2. J Immunol 166:2444–2450
Hirschfeld M, Ma Y, Weis JH, Vogel SN, Weis JJ (2000) Cutting edge: repurification of lipopolysaccharide eliminates signaling through both human and murine toll-like receptor 2. J Immunol 165:618–622
Hoebe K, Janssen E, Beutler B (2004) The interface between innate and adaptive immunity. Nat Immunol 5:971–974
Hoshino K, Takeuchi O, Kawai T, Sanjo H, Ogawa T, Takeda Y, Takeda K, Akira S (1999) Cutting edge: Toll-like receptor 4 (TLR4)-deficient mice are hyporesponsive to lipopolysaccharide: evidence for TLR4 as the Lps gene product. J Immunol 162:3749–3752
Hosoi T, Suzuki S, Nomura J, Ono A, Okuma Y, Akira S, Nomura Y (2004) Bacterial DNA induced iNOS expression through MyD88-p38 MAP kinase in mouse primary cultured glial cells. Brain Res Mol Brain Res 124:159–164

Iliev AI, Stringaris AK, Nau R, Neumann H (2004) Neuronal injury mediated via stimulation of microglial toll-like receptor-9 (TLR9). Faseb J 18:412–414

Jung DY, Lee H, Jung BY, Ock J, Lee MS, Lee WH, Suk K (2005) TLR4, but not TLR2, signals autoregulatory apoptosis of cultured microglia: a critical role of IFN-beta as a decision maker. J Immunol 174:6467–6476

Kaisho T, Akira S (2004) Pleiotropic function of Toll-like receptors. Microbes Infect 6:1388–1394

Kielian T (2006) Toll-like receptors in central nervous system glial inflammation and homeostasis. J Neurosci Res 83:711–730

Kielian T, Bearden ED, Baldwin AC, Esen N (2004) IL-1 and TNF-alpha play a pivotal role in the host immune response in a mouse model of *Staphylococcus aureus*-induced experimental brain abscess. J Neuropathol Exp Neurol 63:381–396

Kielian T, Esen N, Bearden ED (2005) Toll-like receptor 2 (TLR2) is pivotal for recognition of *S. aureus* peptidoglycan but not intact bacteria by microglia. Glia 49:567–576

Kielian T, Mayes P, Kielian M (2002) Characterization of microglial responses to *Staphylococcus aureus*: effects on cytokine, costimulatory molecule, and Toll-like receptor expression. J Neuroimmunol 130:86–99

Kigerl KA, Lai W, Rivest S, Hart RP, Satoskar AR, Popovich PG (2007) Toll-like receptor (TLR)-2 and TLR-4 regulate inflammation, gliosis, and myelin sparing after spinal cord injury. J Neurochem 102:37–50

Laflamme N, Echchannaoui H, Landmann R, Rivest S (2003) Cooperation between toll-like receptor 2 and 4 in the brain of mice challenged with cell wall components derived from gram-negative and gram-positive bacteria. Eur J Immunol 33:1127–1138

Laflamme N, Rivest S (2001) Toll-like receptor 4: the missing link of the cerebral innate immune response triggered by circulating gram-negative bacterial cell wall components. FASEB J 15:155–163

Laflamme N, Soucy G, Rivest S (2001) Circulating cell wall components derived from gram-negative, not gram-positive, bacteria cause a profound induction of the gene-encoding Toll-like receptor 2 in the CNS. J Neurochem 79:648–657

Lehnardt S, Lachance C, Patrizi S, Lefebvre S, Follett PL, Jensen FE, Rosenberg PA, Volpe JJ, Vartanian T (2002) The toll-like receptor TLR4 is necessary for lipopolysaccharide-induced oligodendrocyte injury in the CNS. J Neurosci 22:2478–2486

Lehnardt S, Massillon L, Follett P, Jensen FE, Ratan R, Rosenberg PA, Volpe JJ, Vartanian T (2003) Activation of innate immunity in the CNS triggers neurodegeneration through a Toll-like receptor 4-dependent pathway. Proc Natl Acad Sci USA 100:8514–8519

Lemaitre B, Nicolas E, Michaut L, Reichhart JM, Hoffmann JA (1996) The dorsoventral regulatory gene cassette spatzle/Toll/cactus controls the potent antifungal response in Drosophila adults. Cell 86:973–983

Lien E, Means TK, Heine H, Yoshimura A, Kusumoto S, Fukase K, Fenton MJ, Oikawa M, Qureshi N, Monks B, Finberg RW, Ingalls RR, Golenbock DT (2000) Toll-like receptor 4 imparts ligand-specific recognition of bacterial lipopolysaccharide. J Clin Invest 105:497–504

Ma Y, Haynes RL, Sidman RL, Vartanian T (2007) TLR8: an innate immune receptor in brain, neurons and axons. Cell Cycle 6:2859–2868

Ma Y, Li J, Chiu I, Wang Y, Sloane JA, Lu J, Kosaras B, Sidman RL, Volpe JJ, Vartanian T (2006) Toll-like receptor 8 functions as a negative regulator of neurite outgrowth and inducer of neuronal apoptosis. J Cell Biol 175:209–215

Manukyan M, Triantafilou K, Triantafilou M, Mackie A, Nilsen N, Espevik T, Wiesmuller KH, Ulmer AJ, Heine H (2005) Binding of lipopeptide to CD14 induces physical proximity of CD14, TLR2 and TLR1. Eur J Immunol 35:911–921

Matzinger P (2002) The danger model: a renewed sense of self. Science 296:301–305

Medzhitov R, Janeway C Jr (2000) Innate immunity. N Engl J Med 343:338–344

Medzhitov R, Preston-Hurlburt P, Janeway CA Jr (1997) A human homologue of the Drosophila Toll protein signals activation of adaptive immunity. Nature 388:394–397

Mishra BB, Mishra PK, Teale JM (2006) Expression and distribution of Toll-like receptors in the brain during murine neurocysticercosis. J Neuroimmunol 181:46–56

Mukhopadhyay S, Herre J, Brown GD, Gordon S (2004) The potential for Toll-like receptors to collaborate with other innate immune receptors. Immunology 112:521–530

Nadeau S, Rivest S (2000) Role of microglial-derived tumor necrosis factor in mediating CD14 transcription and nuclear factor kappa B activity in the brain during endotoxemia. J Neurosci 20:3456–3468

Olson JK, Miller SD (2004) Microglia initiate central nervous system innate and adaptive immune responses through multiple TLRs. J Immunol 173:3916–3924

O'Neill LA (2004) TLRs: Professor Mechnikov, sit on your hat. Trends Immunol 25:687–693

O'Neill LA, Bowie AG (2007) The family of five: TIR-domain-containing adaptors in Toll-like receptor signalling. Nat Rev Immunol 7:353–364

Owens T (2005) Toll-like receptors on astrocytes: patterning for immunity. J Neuroimmunol 159:1–2

Palsson-McDermott EM, O'Neill LA (2004) Signal transduction by the lipopolysaccharide receptor, Toll-like receptor-4. Immunology 113:153–162

Pasare C, Medzhitov R (2003) Toll pathway-dependent blockade of CD4 + CD25 + T cell-mediated suppression by dendritic cells. Science 299:1033–1036

Pasare C, Medzhitov R (2004) Toll-like receptors: linking innate and adaptive immunity. Microbes Infect 6:1382–1387

Poltorak A, He X, Smirnova I, Liu MY, Van Huffel C, Du X, Birdwell D, Alejos E, Silva M, Galanos C, Freudenberg M, Ricciardi-Castagnoli P, Layton B, Beutler B (1998a) Defective LPS signaling in C3H/HeJ and C57BL/10ScCr mice: mutations in Tlr4 gene. Science 282:2085–2088

Poltorak A, Smirnova I, He X, Liu MY, Van Huffel C, McNally O, Birdwell D, Alejos E, Silva M, Du X, Thompson P, Chan EK, Ledesma J, Roe B, Clifton S, Vogel SN, Beutler B (1998b) Genetic and physical mapping of the Lps locus: identification of the toll-4 receptor as a candidate gene in the critical region. Blood Cells Mol Dis 24:340–355

Qureshi ST, Lariviere L, Leveque G, Clermont S, Moore KJ, Gros P, Malo D (1999) Endotoxin-tolerant mice have mutations in Toll-like receptor 4 (Tlr4). J Exp Med 189:615–625

Qureshi ST, Medzhitov R (2003) Toll-like receptors and their role in experimental models of microbial infection. Genes Immun 4:87–94

Rasley A, Anguita J, Marriott I (2002) Borrelia burgdorferi induces inflammatory mediator production by murine microglia. J Neuroimmunol 130:22–31

Rivest S (2003) Molecular insights on the cerebral innate immune system. Brain Behav Immun 17:13–19

Saito S, Matsuura M, Tominaga K, Kirikae T, Nakano M (2000) Important role of membrane-associated CD14 in the induction of IFN-beta and subsequent nitric oxide production by murine macrophages in response to bacterial lipopolysaccharide. Eur J Biochem 267:37–45

Saura J (2007) Microglial cells in astroglial cultures: a cautionary note. J Neuroinflammation 4:26

Schroder NW, Morath S, Alexander C, Hamann L, Hartung T, Zahringer U, Gobel UB, Weber JR, Schumann RR (2003) Lipoteichoic acid (LTA) of *Streptococcus pneumoniae* and *Staphylococcus aureus* activates immune cells via Toll-like receptor (TLR)-2, lipopolysaccharide-binding protein (LBP), and CD14, whereas TLR-4 and MD-2 are not involved. J Biol Chem 278:15587–15594

Scumpia PO, Kelly KM, Reeves WH, Stevens BR (2005) Double-stranded RNA signals antiviral and inflammatory programs and dysfunctional glutamate transport in TLR3-expressing astrocytes. Glia 52:153–162

Takeda K, Kaisho T, Akira S (2003) Toll-like receptors. Annu Rev Immunol 21:335–376

Takeshita S, Takeshita F, Haddad DE, Janabi N, Klinman DM (2001) Activation of microglia and astrocytes by CpG oligodeoxynucleotides. Neuroreport 12:3029–3032

Takeuchi O, Hoshino K, Kawai T, Sanjo H, Takada H, Ogawa T, Takeda K, Akira S (1999) Differential roles of TLR2 and TLR4 in recognition of gram-negative and gram-positive bacterial cell wall components. Immunity 11:443–451

Tapping RI, Akashi S, Miyake K, Godowski PJ, Tobias PS (2000) Toll-like receptor 4, but not toll-like receptor 2, is a signaling receptor for *Escherichia* and *Salmonella* lipopolysaccharides. J Immunol 165:5780–5787

Tsan MF, Gao B (2004) Endogenous ligands of Toll-like receptors. J Leukoc Biol 76:514–519

Underhill DM, Gantner B (2004) Integration of Toll-like receptor and phagocytic signaling for tailored immunity. Microbes Infect 6:1368–1373

Weber JR, Freyer D, Alexander C, Schroder NW, Reiss A, Kuster C, Pfeil D, Tuomanen EI, Schumann RR (2003) Recognition of pneumococcal peptidoglycan: an expanded, pivotal role for LPS binding protein. Immunity 19:269–279

Williams MJ, Rodriguez A, Kimbrell DA, Eldon ED (1997) The 18-wheeler mutation reveals complex antibacterial gene regulation in *Drosophila* host defense. EMBO J 16:6120–6130

Yamamoto M, Sato S, Hemmi H, Uematsu S, Hoshino K, Kaisho T, Takeuchi O, Takeda K, Akira S (2003) TRAM is specifically involved in the Toll-like receptor 4-mediated MyD88-independent signaling pathway. Nat Immunol 4:1144–1150

Yamamoto M, Takeda K, Akira S (2004) TIR domain-containing adaptors define the specificity of TLR signaling. Mol Immunol 40:861–868

Zekki H, Feinstein DL, Rivest S (2002) The clinical course of experimental autoimmune encephalomyelitis is associated with a profound and sustained transcriptional activation of the genes encoding toll-like receptor 2 and CD14 in the mouse CNS. Brain Pathol 12:308–319

Zhang Z, Guo K, Schluesener HJ (2005) The immunostimulatory activity of CpG oligonucleotides on microglial N9 cells is affected by a polyguanosine motif. J Neuroimmunol 161:68–77

Zwijnenburg PJ, van der Poll T, Florquin S, Akira S, Takeda K, Roord JJ, van Furth AM (2003a) Interleukin-18 gene-deficient mice show enhanced defense and reduced inflammation during pneumococcal meningitis. J Neuroimmunol 138:31–37

Zwijnenburg PJ, van der Poll T, Florquin S, Roord JJ, Van Furth AM (2003b) IL-1 receptor type 1 gene-deficient mice demonstrate an impaired host defense against pneumococcal meningitis. J Immunol 170:4724–4730

Toll-Like Receptors in Bacterial Meningitis

Uwe Koedel

Contents

Abstract Bacterial meningitis is still an important infectious disease with a high morbidity and mortality rate. Bacterial infection of the cerebrospinal fluid (CSF) space causes a powerful inflammatory reaction that is largely responsibly for meningitis-induced tissue damage and adverse outcome of the disease. In a landmark series of experiments in the mid-1980s, cell wall components including lipooligosaccharides and lipoteichoic acid were indicated to be the key bacterial elements that can trigger the host inflammatory response in the CSF. Ten years ago, the discovery of Toll-like receptor proteins (TLRs) that allow the detection of microbial components and initiate the host immune response opened up new horizons in research on the pathophysiology of meningitis. Cell culture approaches provided the first evidence for a crucial role of TLRs in sensing meningeal pathogens including *Streptococcus pneumoniae*, *Neisseria meningitidis*, *Streptococcus agalactiae*, and *Listeria monocytogenes*. Subsequently, studies in mice with single or combined deficiencies in TLRs demonstrated that TLR activation is a key

U. Koedel
Clinic of the Ludwig-Maximilians, Department of Neurology, University of Munich, Marchioninistr 15, Munich, 81377, Germany
e-mail: Uwe.Koedel@med.uni-muenchen.de

T. Kielian (ed.), *Toll-like Receptors: Roles in Infection and Neuropathology*, Current Topics in Microbiology and Immunology 336,
DOI 10.1007/978-3-642-00549-7_2,

event in meningeal inflammation and, even more interestingly, a pivotal factor for meningitis-associated tissue damage. A detailed understanding of the mechanisms of host–pathogen interactions in the CSF space may generate new opportunities for specific treatment strategies for bacterial meningitis.

Abbreviations

BCG	Bacillus Calmette-Guérin
cfu	Colony-forming unit
CHO	Chinese hamster ovary
CNS	Central nervous system
CR	Complement receptor
CSF	Cerebrospinal fluid
DAI	DNA-dependent activator of interferon regulatory factors
GBS	Group B streptococci
HEK	Human embryonic epithelial kidney
HIB	*Haemophilus influenzae* type B
IL	Interleukin
ie-DAP	Meso-diaminopimelic acid
ISH	In situ hybridization
LPS	Lipopolysaccharide
LTA	Lipoteichoic acid
MDP	Muramyl dipeptide
MR	Mannose receptor ·
MyD	Myeloid differentiation factor
NALP3	Nacht domain, leucine-rich repeat and PYD-containing protein 3
NOD	Nucleotide-binding oligomerization domain
NLR	NOD-like receptor
PAMP	Pathogen-associated molecular pattern
PCV	Pneumococcal conjugate vaccine
PGN	Peptidoglycan
PPR	Pattern recognition receptor
RIP	Protein kinase receptor-interacting protein
SR	Scavenger receptor
TLR	Toll-like receptor
TNF	Tumor necrosis factor

1 Introduction

From its first description by Vieusseux in 1806 until the early twentieth century, acute bacterial meningitis was quite uniformly considered a fatal disease. Although the introduction of antibiotics made it curable, mortality and morbidity from the disease remain unacceptably high (Van de Beek et al. 2004). While endemic bacterial meningitis remains relatively rare, particularly in developed countries, the potential occurrence of epidemic bacterial meningitis in any part of the world heightens its profile as a major infectious disease. Large epidemic waves of meningococcal meningitis occur with a periodicity of 8–12 years in sub-Saharan territories of Africa, the so-called meningitis belt. In 1996, Africa experienced the largest recorded outbreak of epidemic meningitis in history, with over 250,000 cases and 25,000 deaths. Over 35,000 of the cured patients are expected to have sustained permanent neurological, otological, and psychological deficits (Robbins et al. 1997). In addition to these epidemics, at least 1.2 million cases of bacterial meningitis are estimated to occur each year; 135,000 of them are fatal. These numbers have made bacterial meningitis one of the top ten infectious causes of death worldwide. Clinical and neuropathological studies have clearly indicated that fatal disease outcome is predominantly caused by intracranial complications, including raised intracranial pressure, brain edema formation, and cerebrovascular insults (Kastenbauer and Pfister 2003; Van de Beek et al. 2004). During the past two decades, the frontier of investigation has therefore focused on the pathophysiology of meningitis-associated brain injury. It became evident that the host immune response to the pathogen, rather than the pathogen itself, is largely responsible for the damage that results from bacterial meningitis (Koedel et al. 2002a; Weber and Tuomanen 2007). As a result, the therapeutic approach to bacterial meningitis has to be widened from eradicating the pathogen with antibiotics to preventing the detrimental effects of the host immune response. The intention of this chapter is to provide an overview of the current state of our knowledge in this field by addressing the following questions: what do we know about the immune milieu at the site of infection? Which pathogens are the major causes of purulent meningitis? How are these bacteria recognized by immune cells? What is the impact of these "recognition systems" on the clinical course of the disease?

2 An Immunological Approach to the Cerebrospinal Fluid Compartment

Bacterial meningitis is mainly an acute purulent infection of the leptomeninges and subarachnoid space caused by bacteria. The leptomeninges are the two innermost layers of tissue (the arachnoid mater and pia mater) that envelop the brain. The subarachnoid space is the interval between the leptomeninges, which is filled with cerebrospinal fluid (CSF). The meninges and the CSF serve to cushion the brain, thereby protecting it from injury.

From an immunological point of view, the subarachnoid space is a unique site in the body (Fig. 1). Specialized blood–CSF barriers seclude the subarachnoid space from the circulating blood and prevent most blood components from entering the CSF (Pachter et al. 2003). Moreover, soluble pattern recognition receptors (PRR, e.g., complement factors) that recognize bacteria and enhance their uptake

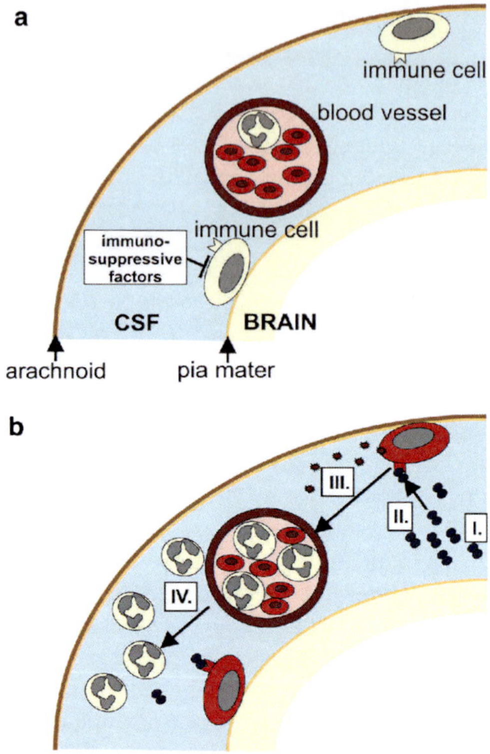

Fig. 1a–b Schematic illustration of the (**a**) normal and (**b**) inflamed subarachnoid space. **a** From an immunological point of view, the subarachnoid space is a unique anatomical compartment. It shares several features with immune-privileged organs. For example, soluble pattern recognition receptors that recognize bacteria and enhance their uptake by phagocytes are largely absent in the CSF. Moreover, the normal CSF contains an array of immunosuppressive factors that actively suppress immune reactivity. However, contrary to immune-privileged organs, functionally active immune cells are present in all non-neural structures that are in direct contact with the CSF. These cells can function as sentinel cells that are able to detect the presence of bacteria in the CSF. **b** Once pathogens have entered the subarachnoid space, they take advantage of the immunodeficiency within the CSF and multiply easily and efficiently (*I*). Phagocytes recognize pathogens by means of their cell surface receptors, so-called pattern recognition receptors (*II*). The activated phagocytes can destroy pathogens without any additional help, but in almost all cases they amplify immunity by releasing cytokines and chemokines (*III*). As a consequence, large numbers of neutrophils are recruited into the CSF (*IV*), a pathologic hallmark of acute bacterial meningitis

by phagocytes are largely absent in the CSF (Dujardin et al. 1985; Stahel et al. 1997). Additionally, the subarachnoid cavity lacks fully organized drainage via lymphatic vessels (Johnston et al. 2004). Lymphatic vessels are of crucial importance for the migration of antigen-presenting dendritic cells from tissues to lymph nodes, where they interact with T and B lymphocytes to mount and shape an adaptive immune response (Steinman 2007). Finally, the normal CSF contains an array of anti-inflammatory and immunosuppressive factors that actively suppress immune reactivity (Niederkorn 2006).

However, in contrast to the brain, which lacks competent antigen-presenting cells (microglia, the resident cells for immune defenses in the brain, are in a resting state under physiological conditions, with weak expression of molecules associated with antigen presentation; Schwartz et al. 2006), functionally active macrophages and dendritic cells are present in all non-neural structures that are in direct contact with the CSF, namely the leptomeninges, the perivascular spaces and the choroid plexus (Pashenkov and Link 2002; Guillemin and Brew 2004). These cell types can function as sentinel cells to detect the presence of bacteria in the CSF through PRRs. Gene transcription and protein analyses have verified that diverse PRRs are expressed in tissues lining the CSF compartment. For instance, macrophage scavenger receptors (SR) that can bind and internalize a variety of microbial pathogens (Mukhopadhyay and Gordon 2004) are found on stromal and epiplexus macrophages of the choroid plexus, meningeal macrophages and on perivascular sites, but not on brain microglia (Naito et al. 1991). Likewise, the mannose receptor (MR) that recognizes a range of carbohydrates present on microbes and mediates microbial phagocytosis (McGreal et al. 2005) is exclusively expressed by perivascular, meningeal, and choroid plexus macrophages in the brain (Galea et al. 2005). The phagocytic complement receptors like CR1 (Roozendaal and Carroll 2006) exhibit a brain tissue expression pattern identical to that of SRs and MR (Singhrao et al. 1999). In addition, signaling PRRs, including Toll-like receptors (TLR) and nucleotide-binding oligomerization domain (NOD) proteins, have been implicated as being present on and in immunocompetent cells within the brain (Bsibsi et al. 2002; Letiembre et al. 2007; Rodriguez-Martinez et al. 2005). In situ hybridization (ISH) studies revealed a high constitutive expression of TLR2 and TLR4 mRNA in the choroid plexus. A strong hybridization signal for TLR4 was also observed along leptomeningeal vessels (Laflamme and Rivest 2001; Laflamme et al. 2001, 2003). The brain expression of other TLRs (e.g., TLR8 and TLR9) was below the detection limit of the ISH technique (De Luigi et al. 1998). Collectively, the normal milieu in the CSF seems to be much more "immune friendly" compared to that of the intact brain parenchyma, which constitutes a milieu that is unfavorable for immune responses. The difference in immune reactivity between both CNS compartments is exemplarily illustrated by experiments in which heat-killed bacillus Calmette–Guérin (BCG) was injected into either site (Matyszak and Perry 1996). Intra-CSF administration induced extensive leukocyte infiltration within the first few hours, comparable to that in immunocompetent tissues like the skin, whereas the same dose of BCG evoked only a minimal inflammatory response in the brain parenchyma (Matyszak and Perry 1996). However, in contrast to immunocompetent tissues, the host immune response within the CSF compartment is largely inefficient at controlling and overcoming microbial infections. For instance, *Streptococcus pneumoniae*, a major meningeal pathogen, can multiply within the CSF almost as efficiently as it can in vitro, reaching high titers of up to 10^9 colony-forming units (cfu)/ml. Moreover, intracisternal inoculation of 10^4 cfu of *S. pneumoniae* in mice resulted in the death of all infected animals, whereas only one quarter of mice died after intratracheal application of an identical dose of this pneumococcal strain (Gerber et al. 2001; Rijneveld et al. 2001). Similar data have also been reported for humans. In the preantibiotic era, pneumococcal meningitis was a lethal disease, while 75% of patients survived pneumococcal pneumonia (Heffron 1979). This local host defense deficiency seems to be due to a local absence of soluble PRRs that flag the pathogens for phagocytosis, the local expression of immunosuppressive factors that interfere with the uptake and killing of pathogens, or both. The failure to eliminate the pathogens is the "driving force" for the immune response that is a key mediator of meningitis-associated tissue injury, and is thus largely responsible for the adverse disease outcome.

3 Bacteriologic Profile of Community-Acquired Bacterial Meningitis

During the past decades, we have witnessed significant changes in the epidemiology of bacterial meningitis. When all cases occurring after the neonatal period are taken into account, the leading cause of bacterial meningitis 20–30 years ago was *Haemophilus influenzae* type B (HiB), with a relative frequency of about 45%, followed by *S. pneumoniae* with 18% and *Neisseria meningitidis* with 14% (Swartz 2004). The introduction of the HiB conjugate vaccine in the early 1990s has virtually eliminated invasive HiB disease in countries that have included the HiB vaccine in their national immunization programs. By the end of 2005, HiB vaccines were part of the routine infant immunization program in 101 of the 192 WHO member states. Unfortunately, the proportion of infants in the world who had benefited from the HiB vaccination by the end of 2005 was still fairly low (21%), as eight countries with large birth cohorts had not introduced the vaccine (Rossi et al. 2007). In countries with HiB vaccination programs, *S. pneumoniae* became the leading species (with a total of 47% cases in the US), followed by *N. meningitidis* (25%), group B streptococci (12%) and *Listeria monocytogenes* (8%) (Schuchat et al. 1997). Further breakthroughs were the introduction of pneumococcal and meningococcal conjugate vaccines in the early 2000s (Makwana and Riordan 2007). For instance, the pneumococcal conjugate vaccine (PCV7) protects against the seven most common pneumococcal strains (out of over 90 serotypes) that cause invasive disease, including bloodstream infections and meningitis. After the implementation of routine childhood vaccination with PCV7 in the US, the average hospitalization rates for pneumococcal meningitis decreased by about 66% among children younger than 2 years and by 33% among adults aged 65 years and older (Tsai et al. 2008). The decline in invasive diseases in older adults probably occurs because of a decreased community transmission of vaccine-type pneumococci from young children to adults. Admittedly, the effectiveness of the vaccine against all invasive diseases, without regard to serotype, could be diminished in that nonvaccine pneumococcal serotypes account for larger portions of the disease (Whitney et al. 2006). All in all, the introduction of these bacterial vaccines has led to a decrease in the rate of bacterial meningitis (in countries with the power to or the support of the international community necessary to implement these costly programs), and bacterial meningitis is now a disease predominantly found in adults rather than in infants and young children. Meningitis due to *N. meningitidis* is the most common form among young adults, whereas meningitis due to *S. pneumoniae* meningitis dominates in older adults (Dery and Hasbun 2007; Schut et al. 2008) (Table 1). These two pathogens account for over 80% of all cases of bacterial meningitis. Other etiologic agents include *Streptococcus agalactiae*, *Escherichia coli*, *L. monocytogenes*, and *Mycobacterium tuberculosis* (Schuchat et al. 1997; Van de Beek et al. 2004). *S. agalactiae* and *E. coli* are the leading causes of meningitis among newborns (Heath et al. 2003). *L. monocytogenes* is an uncommon cause of meningitis in the general population, but an important pathogen in neonates, the elderly, transplant recipients and other patients with impaired cell-mediated immunity (Doganay 2003). *M. tuberculosis* very rarely causes meningitis in Western countries but is feared in countries where tuberculosis is endemic (Thwaites et al. 2000).

Bacterial meningitis typically begins with host acquisition of the abovementioned pathogens by nasopharyngeal colonization, followed by systemic invasion, development of high-grade bacteremia, and invasion of the subarachnoid space via the blood–brain/CSF barrier. Pathogens can also enter the subarachnoid space by direct migration from nearby infections (e.g., sinusitis, mastoiditis) or through exterior openings in normally closed CSF pathways (e.g., due to meningomyelocele or neurosurgical procedures). Once the pathogens have entered the subarachnoid space, they take advantage of the immunodeficient CSF compartment and multiply easily and efficiently. The host immune response does not appear to be activated until the pathogens reach a relatively high density (i.e., $>10^5$ cfu/ml for *S. pneumoniae*) (Tuomanen et al. 1985b). During recent decades, the mechanisms underlying host–pathogen recognition and immune activation have been intensively investigated using in vitro and in vivo approaches. The following sections will address how TLRs contribute to the recognition of the most common pathogens that cause acute bacterial meningitis in developed countries, namely *S. pneumoniae*, *N. meningitidis*, *S. agalactiae*, and *L. monocytogenes*.

4 Host Cellular Sensors of Meningeal Pathogens

In a landmark series of studies on immune activation in pneumococcal meningitis (Tuomanen et al. 1985a,b, 1986), the cell wall was found to be the key bacterial element that can trigger the host inflammatory response. Both of the major pneumococcal cell wall components, lipoteichoic acid (LTA) and peptidoglycan (PGN), can contribute to the induction of meningeal inflammation. Each has specific activity that is high enough to produce an immune response if given at doses of $\geq 10^5$ bacterial equivalents (Tuomanen et al. 1985a). Intracisternal inoculation of the cell wall induces clinical symptoms of meningitis (Tuomanen et al. 1989). The greater the amount of cell wall components in the CSF, the worse the clinical outcome of

Table 1 Epidemiologic data for acute bacterial meningitis

Pathogen	Age group	Frequency			Case fatality rate		
		Netherlands	France	Central African Republic	Netherlands	France	Central African Republic
		Adults (*n* = 696)	Children (*n* = 1,084)	Children (*n* = 167)	Adults (*n* = 696)	Children (*n* = 1,084)	Children (*n* = 167)
S. pneumoniae	Neonates Children Adults	51%	33%	37%	30%	11%	47%
N. meningitidis	Children Adults	37%	55%	5%	7%	8%	13%
S. agalactiae	Neonates	0.7%	5%	–[a]	–[b]	19%	
L. monocytogenes	Neonates Adults	4%	0.5%	–[a]	–[b]	17%	
Others: *H. influenzae*	Children	2%	2.5%	28%	–[b]	4%	33%

The occurrence and outcome of acute bacterial meningitis varies according to socioeconomic aspects (developed vs. developing countries), age, and the pathogen causing the acute bacterial meningitis. Data are taken from van de Beek et al. (2004); Bingen et al. (2005) Bercion et al. (2008)

[a] Not detected

[b] Data not reported

the disease (Schneider et al. 1999). When challenged with pneumococcal cell wall components, isolated immunocompetent cells produce a wide range of cytokines, like interleukin (IL)-1β, IL-6, tumor necrosis factor (TNF)-α, CXCL1, or CXCL2 (Riesenfeld-Orn et al. 1989; Heumann et al. 1994; Hanisch et al. 2001). Similar to this Gram-positive pathogen, the inflammatory activity of the Gram-negative bacteria *N. meningitidis* and HiB was mainly attributed to the release of bacterial cell wall material. Thus, in rabbits, intra-CSF injection of HiB lipooligosaccharide (LOS, but not capsular polysaccharide) induces an inflammatory reaction in the subarachnoid space comparable to that of live HiB (Syrogiannopoulos et al. 1988). Likewise, injection of meningococcal lipopolysaccharide (LPS) into the CSF compartment results in meningitis (Waage et al. 1989). In patients with meningococcal disease, the compartmentalization of LPS production correlates with the clinical presentation, meaning that high LPS levels in CSF and low LPS in plasma are associated with meningitis, whereas reciprocal findings are obtained from patients with septicemia (Brandtzaeg et al. 1992). Collectively, it is virtually certain that components of the bacterial cell walls "can do it all" when it comes to initiating the host immune response in meningitis. The essential question, however, concerns how bacterial cell wall components can trigger host responses.

Two studies in the mid-1990s solved the mystery of pathogen recognition by the immune system. One study showed that signaling through the mammalian receptor TLR4 activates the immune system (Medzhitov et al. 1997). The other study found that TLR4 plays a critical role in mediating the inflammatory activity of the Gram-negative cell wall component LPS (Poltorak et al. 1998). These discoveries were foreshadowed by the observation that fruit flies require the protein Toll, the namesake of the TLR family, to sense fungal infection (Lemaitre et al. 1996). The fact that a single receptor family serves as a sentinel of microbial infection in such widely different species suggested a fundamental role of this recognition mechanism. It soon became clear that the recognition of microbial invaders is based on the detection of conserved microbial molecules (pathogen-associated molecular patterns or PAMPs) (Medzhitov and Janeway 2002; Beutler and Rietschel 2003; Akira et al. 2006) that are (1) produced only by microbes, (2) invariant between microorganisms of a given class, and (3) essential for microbial survival. These microbial molecules are recognized by specific receptors of the innate immune system, the PRRs. A recently published database of PRR (called PRRDB) contains around 500 PRRs from 77 distinct organisms ranging from insects to humans. This includes 177 TLRs, 124 SRs, and 67 NOD-like receptors (NLRs) (Lata and Raghava 2008). A total of 13 TLRs have been identified in mammals; humans express TLR1–10, while mice express TLR1–7, TLR9 and TLR11–13 (Beutler et al. 2006). To date, TLRs are the best-characterized class of PRRs. These transmembrane receptors are expressed on the cell surface or intracellularly on endosomal membranes (TLR3, 7, 8, and 9), and binding of the respective ligands elicits inflammatory and antimicrobial responses of the host. The first insights into the role of TLRs as sensors of major meningeal pathogens were provided by cell culture studies (Fig. 2).

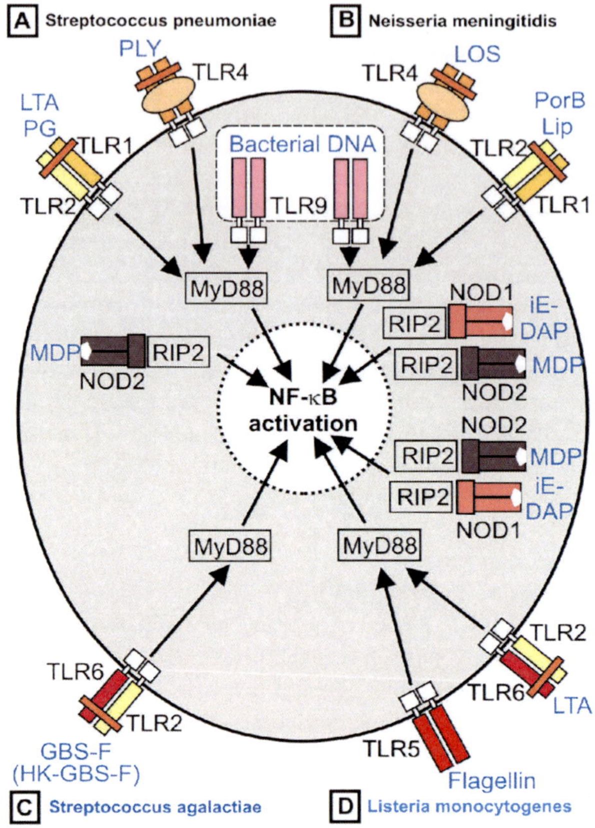

Fig. 2a–d Model for engagement of TLRs and NODs by major meningeal pathogens, as deduced from cell culture experiments. *S. pneumoniae* (**a**) generally has the potential to activate cells through TLR1/2, TLR4, TLR9, NOD2, and presumably as-yet unidentified pattern recognition receptors. Pneumococcal peptidoglycan (PGN) and lipoteichoic acid (LTA) are recognized through TLR1/2, whereas TLR4 senses the presence of pneumolysin (PLY). Moreover, pneumococcal DNA and internalized muramyl dipeptide (MDP) are detected through TLR9 and NOD2, respectively. A similar subset of pattern recognition receptors has been implicated in the host recognition of *Neisseria meningitidis* (**b**). While meningococcal lipooligosoccharide (LOS) and DNA interact with TLR4 and TLR9, respectively, the meningococcal outer membrane proteins porin B (PorB) and Lip are recognized through TLR2 with the help of TLR1 and CD14. In addition to MDP, the meningococcal cell wall contains dipeptide meso-diaminopimelic acid (iE-DPP), which induces activation of the transcription factor NF-κB through NOD1-RIP2-dependent signaling. In contrast to these pathogens, the mechansims involved in the sensing of *S. agalactiae* (**c**) are still largely unclear. Recently, a heat-labile factor named GBS-F has been found to engage the TLR2/TLR6 heterodimer as an essential signaling molecule. Similar to *N. meningitidis*, *L. monocytogenes* (**d**) has been found to express NOD1- and NOD2-dependent immunostimulatory activity in vitro. Extracellular listerial cell wall components such as LTA are recognized by TLR2 (with the help of CD14 and TLR6), whereas extracellular listerial flagellin signals through TLR5. Other TLRs including TLR4 and TLR9 have been reported as being dispensable for *Listeria*-induced immune cell activation

4.1 Immune Recognition of S. pneumoniae *In Vitro*

In 1999, Yoshimura and colleagues (Yoshimura et al. 1999) reported that heterologous expression of human TLR2, but not TLR4, in Chinese hamster ovary (CHO) fibroblasts confers responsiveness to heat-killed *S. pneumoniae* (D39, a serotype 2 strain), as evidenced by inducible translocation of the transcription factor NF-κB. The role of TLR2 in the sensing of *S. pneumoniae* was further strengthened by the following findings: first, the human embryonic epithelial kidney cell line HEK293 becomes responsive to heat-killed (D39 or TIGR4, a serotype 4 strain), antibiotic-lysed encapsulated (serotype 3 strain), or live unencapsulated pathogen (strain R6x) when transfected with TLR2 (Malley et al. 2003; Koedel et al. 2003; Schmeck et al. 2006). Second, the uptake and killing of pneumococci (serotype 1 and 3 strains) by isolated murine polymorphonuclear leukocytes is impaired in the absence of TLR2 (Letiembre et al. 2005). The cell wall components, PGN and LTA were identified as the main pneumococcal ligands for TLR2 (Yoshimura et al. 1999; Schroder et al. 2003). TLR2 was reported to recognize pneumococcal LTA in cooperation with TLR1 (Han et al. 2003) and with the help of CD14 (Schroder et al. 2003; Han et al. 2003). In addition to TLR2, other TLRs may also contribute to sensing the presence of *S. pneumoniae*. For instance, Malley and colleagues (Malley et al. 2003) demonstrated that ethanol killed *S. pneumoniae* (D39 strain) stimulated HEK293 cells transfected either with TLR2 or TLR4. Similarily, antibiotic-lysed pneumococci (serotype 3 strain) were found to trigger activation of transfected HEK293 cells in a TLR2- and TLR4-dependent manner (Koedel et al. 2003). The observations that ethanol-killed pneumolysin-deficient pneumococci (isogenic mutant of D39) failed to stimulate TLR4-expressing HEK293 cells, and that recombinant pneumolysin failed to activate TLR4-deficient macrophages but was effective on macrophages from TLR4-positive mice, suggested that TLR4 mediates host immune responses to pneumococci through its interaction with one of the most important virulence factors of the organism, the cholesterol-dependent cytolysin pneumolysin (Malley et al. 2003; Schmeck et al. 2006). Interestingly, TLR4 seems to be dispensable for the cellular recognition of viable *S. pneumoniae* (Mogensen et al. 2006; Schmeck et al. 2006). Instead, TLR9 was implicated as another sensor for the live bacterium. Mogensen and colleagues (Mogensen et al. 2006) reported that viable *S. pneumoniae* (TIGR4 strain) induced cell activation in HEK293 cells that were stably transfected with TLR9. TLR9-dependent cell activation was also observed when isolated genomic DNA from *S. pneumoniae* was used as a stimulus (Mogensen et al. 2006). In contrast to viable pneumococci and their genomic DNA, heat-killed (serotype 4 and 14 strain) and antibiotic-lysed pneumococci (serotype 3 strain) were almost completely unable to activate TLR9 (Koedel et al. 2003; Mogensen et al. 2006; Lee et al. 2007). The reported differences in the role of single TLRs in pneumococcal sensing are most likely attributable to the fact that distinct pneumococcal strains and preparations express and/or release distinct PAMPs. For example, heat inactivation of bacteria can lead to unwanted inactivation of PAMPs, as demonstrated for the TLR4 ligand pneumolysin (Malley et al. 2003). The pneumococcal capsule might also interfere with the release of single PAMPs upon the heat killing of pneumococci. This was demonstrated by the intracisternal challenge of animals with heat-killed pneumococci. Nonencapsulated, heat-killed pneumococci elicited an inflammatory response in intracisternally challenged animals, whereas encapsulated heat-killed bacteria did not. In contrast, such a difference was not observed when viable encapsulated and unencapsulated pneumococci were compared (Tuomanen et al. 1985b). Moreover, large disparities have been reported for the production of the TLR4 ligand pneumolysin by different strains. Benton and coworkers described two principal types of pneumolysin production during in vitro growth (Benton et al. 1997). For the D39 (serotype 2) strain, they found a rise of cytoplasmic pneumolysin but no pneumolysin release until the late log phase. In contrast, for the Wu (serotype 3) strain, a measurable extracellular pneumolysin titer became detectable prior to cytoplasmic titers, possibly due to small amounts of autolysis that can begin in the early- to mid-log phase (Benton et al. 1997). Furthermore, sequence variations in the pneumolysin gene were reported to occur at a frequency higher than 3% (Jefferies et al. 2007). Allelic variations can result in the expression of noncytolytic pneumolysin in certain clonal groups of pneumococci, namely lineages of the serotypes 1 and 8 (Jefferies et al. 2007). In a mouse model of bronchopneumonia, mice infected with an isogenic mutant strain that expresses noncytolytic pneumolysin showed a pattern of neutrophil infiltration distinct from that observed in mice infected with the wild-type strain. In the early phase after infection (12 h), the numbers were the same with both strains, but neutrophil numbers were subsequently significantly lower in mice infected with the mutant strain than in animals inoculated with the wild-type strain (Jounblat et al. 2003). The existence of pneumolysin variants among pneumococcal strains suggests that, although the reduced inflammatory activity of noncytolytic pneumolysin is not due to a loss of binding to TLR4 (Malley et al. 2003), differences in PAMP structure between pneumococcal strains may also contribute to the reported differences in the role of single TLRs in pneumococcal sensing. Collectively, in vitro studies have demonstrated that *S. pneumoniae* generally has the potential to activate cells through TLR2, TLR4, and TLR9 (Fig. 2). Utilizing primary immune cells isolated from mice with single, double or triple deficiencies, studies by Snapper's (Lee et al. 2007) and our group (Klein et al. 2008) recently provided evidence that a single deficiency of TLR2, TLR4 or TLR9 caused only selective and relatively modest reductions in cytokine production by pneumococci-stimulated immune cells, whereas the combined loss of TLR2, TLR4 and TLR9 recapitulated the phenotype of cells lacking MyD88, an essential adaptor molecule in the signaling cascade of all TLRs with the exception of TLR3 (Takeuchi and Akira 2002; Barton and Medzhitov 2003). These in vitro data strongly suggest that distinct TLRs must synergize to fully activate an immune cell response to *S. pneumoniae*.

While there is no doubt that TLRs are important for sensing pneumococcal infection, there is also substantial evidence for the involvement of other PRRs in pneumococcal recognition. The NLRs are one example of such a receptor (Kanneganti et al. 2007). Whereas TLRs are able to sense PAMPs at the cell surface and within endosomes, NLRs detect PAMPs in the cytosol. The discovery of these cytosolic PRRs suggests that pathogens evading extracellular surveillance encounter another line of recognition in the cytosol (Kanneganti et al. 2007). Two NLRs, NOD1 and NOD2, sense the cytosolic presence of peptidoglycan fragments and lead to protein kinase receptor interacting protein (RIP)-2-dependent activation of NF-κB. A different set of NLRs (like NALP1, NALP3 or IPAF) are qualified to recognize a plethora of molecules ranging from divergent PAMPs to endogenous danger molecules and promote the assembly of inflammasome complexes (Ogura et al. 2006) that are required for the activation of caspase-1 (and the generation of biologically active IL-1 family cytokines). Recently, Opitz et al. (2004) demonstrated that pneumococci can invade HEK293 cells and induce NF-κB activation through NOD2- and RIP2-dependent signaling. Further studies revealed that NOD2 is activated by muramyl dipeptide (MDP), a component of virtually all types of PGN (Girardin et al. 2003b), whereas NOD1 recognizes the dipeptide meso-diaminopimelic acid (iE-DAP) which is found in many Gram-negative and certain Gram-positive bacteria (Girardin et al. 2003a). Accordingly, many bacteria, including the meningeal pathogens *N. meningitidis* (Girardin et al. 2003a) and *L. monocytogenes* (Hasegawa et al. 2006; Park et al. 2007), have been shown to express NOD1- and/or NOD2-stimulatory activity in vitro. Thus, NOD1 and NOD2 may act as sensors of microbial invaders of the CSF compartment—presumably in cooperation with TLRs (and additional PRRs).

4.2 *Immune Recognition of* N. meningitidis *In Vitro*

TLRs have also been ascribed a particularly important role in the initiation of the host immune response against *N. meningitidis* (Fig. 2). By genetic complementation in HEK293 cells, human TLR2, TLR4, and TLR9 were found to confer responsiveness to live *N. meningitidis* strains (Mogensen et al. 2006). The ability of *N. meningitidis* to activate TLR2 involves recognition of the neisserial porin, the major outer membrane protein of the pathogenic *Neisseria* (Massari et al. 2002). Thereby, TLR2 recognizes the porin PorB through direct binding, and engagement of the TLR2/TLR1 heterodimer is required for initiating signaling in transfected HEK293 cells and in murine B cells (Massari et al. 2006). Apart from neisserial porin, purified Lip lipoprotein, which contains a conserved epitope known as H.8 that is common to all pathogenic *Neisseria* species, is capable of stimulating the production of proinflammatory mediators from HEK293 cells in a TLR2-dependent manner. The Lip-induced activation of TLR2-expressing HEK293 cells is further enhanced by co-transfection of TLR1 (but not TLR6) (Fisette et al. 2003). With respect to TLR4, meningococcal LOS is necessary for receptor activation, as a LOS-deficient *N. meningitidis* mutant was unable to induce cell activation via TLR4/MD2 (Pridmore et al. 2001). TLR4-mediated activation requires engagement of both CD14 and MD-2 (Zughaier et al. 2004), which is typical of the pattern of TLR4 binding by Gram-negative bacteria. By using an isogenic mutant system, truncation of the oligosaccharide and removal of the capsule were found to exert no effect on the ability of meningococcal LOS to signal via the TLR4/MD2 complex (Pridmore et al. 2003). More recent studies indicated that meningococcal KDO_2-lipid (KDO = 3-deoxy-D-manno-octulosonic acid) is the minimal structure required for immunostimulatory activity of meningococcal LOS (Zughaier et al. 2006, 2007). Moreover, natural DNA sequences present in Gram-negative bacteria including *N. meningitidis* were shown to produce innate immune cell stimulation via TLR9 (Magnusson et al. 2007). The involvement of TLR9 in meningococcal detection was further strengthened by the study of Mogensen et al. (2006) that demonstrated the ability of both purified meningococcal DNA and live *Neisseria* to stimulate cells through TLR9. This study also showed that *Neisseria* can activate parental HEK293 cells, devoid of any TLRs, suggesting the contribution of TLR-independent signaling pathways in *Neisseria*-induced cell activation.

4.3 Immune Recognition of **S. agalactiae** *In Vitro*

The first study that investigated the role of TLRs in the recognition of group B streptococci (GBS or *S. agalactiae*) was published in 2000 (Flo et al. 2000) (Fig. 2). This study reported that expression of neither TLR2 nor CD14 makes CHO cells responsive to heat-killed GBS, and that a blocking antibody directed against TLR2 does not inhibit monocyte activation upon exposure to heat-killed GBS. More recent experiments using peritoneal macrophages from MyD88- and TLR2-deficient mice indicated that TLR2 is involved in TNF-α production to heat-labile, extracellular GBS products but not the whole bacterium, whereas the induction of TNF-α by both types of stimuli depends entirely upon MyD88 (Henneke et al. 2001). The heat-labile factor (named "GBS-F") engages the TLR2/TLR6 heterodimer as the essential signaling molecule (Henneke et al. 2001). The requirement of MyD88 to sense the whole heat-inactivated bacterium suggested that GBS is recognized by TLRs (or alternatively by receptors of the IL-1R family) other than TLR2. By using macrophages from mutant mice carrying spontaneous mutations or targeted deletions of individual TLRs, all of the TLRs that have been implicated as sensors of microbial products from Gram-positive bacteria (TLR1, TLR2, TLR4, TLR6, TLR9) could be excluded as being solely responsible for the induction of TNF-α release by heat-killed GBS (Henneke et al. 2001, 2002). However, a recent analysis of gene expression profiles in wild-type and TLR2-deficient macrophages demonstrated that, while the upregulation of 76% of genes (including TNF-α) induced by GBS in macrophages does not depend on TLR2, the induction of a small fraction of genes is impaired in the absence of TLR2 (Draper et al. 2006). Among these genes is the cytokine IL-1β that signals through MyD88. This observation raises the possibility that the abrogation of cytokine production in MyD88-deficient cells in response to heat-killed GBS is, at least partly, due to an interrupted feedback loop via IL-1 family receptors. In addition, the engagement of TLR7 and/or TLR8 by microbial RNA may possibly contribute to MyD88-dependent macrophage activation by GBS.

4.4 *Immune Recognition of* L. monocytogenes *In Vitro*

While the story of GBS-induced immune activation is far from clear, *L. monocytogenes* infection is evidently sensed by diverse cell surface and cytosolic PRRs. Listerial cell wall components such as LTA are recognized by TLR2 (with the help of CD14 and TLR6, Fig. 2) (Flo et al. 2000; Seki et al. 2002; Janot et al. 2008), whereas extracellular listerial flagellin signals through TLR5 (Hayashi et al. 2001). Other TLRs including TLR3, TLR4 and TLR9 have been reported as being dispensable for *Listeria*-induced immune activation (Janot et al. 2008). Once in the cell, live bacteria replicating in the cytosol activate a macrophage transcriptional response distinct from that of *Listeria* trapped in phagosomal vacuoles (Leber et al. 2008). The vacuolar response was found to be entirely MyD88-dependent and to control the induction of many proinflammatory cytokines, including IL-1β and TNF-α (Leber et al. 2008). The cytosolic response was reported to induce a distinct and significantly nonoverlapping set of host response genes including Type 1 interferons, and to be under the coordinated control of at least two pathways. The first pathway consists of recognition of bacterial DNA by cytosolic PRR(s), possibly DNA-dependent activator of IFN regulatory factors (DAI) (Takaoka et al. 2007; Wang et al. 2008). The second pathway involves the detection of listerial PGN components, presumably by NOD1 and NOD2 (Park et al. 2007; Hasegawa et al. 2006). Further studies showed that cellular activation upon *L. monocytogenes* infection critically depends on NOD1 and NOD2 in macrophages pretreated with TLR ligands, but not in naïve macrophages (Kim et al. 2008), and that other receptors of the NLR family including NALP3 and IPAF detect cytosolic *Listeria* and subsequently activate caspase-1 (Mariathasan et al. 2006; Franchi et al. 2007; Warren et al. 2008). These data suggest that multiple NLRs are crucial for microbial recognition and host defense against *L. monocytogenes*, particularly when proinflammatory responses are compromised by tolerization induced by TLR stimulation.

Collectively, in vitro studies indicated that distinct yet overlapping sets of TLRs and other PRRs (namely the NLRs NOD1 and NOD2) are used to sense major meningeal pathogens including *S. pneumoniae*, *S. agalactiae*, *N. meningitidis*, and *L. monocytogenes*. However, in vivo studies using mice with targeted deletions of individual TLRs (and other PRRs) are needed to assess whether these results can be extrapolated to in vivo models of bacterial meningitis. So far, the in vivo relevance of TLRs for mounting an inflammatory response to meningeal infection has been demonstrated only for *S. pneumoniae*, as outlined in the next section.

5 Role of TLRs in Pneumococcal Meningitis

The first evidence for TLR involvement in pneumococcal meningitis was provided by Echchannaoui et al. (2002), who demonstrated that while intracerebral infection (into the left forebrain) with *S. pneumoniae* was fatal in both wild-type and

TLR2-deficient mice, the clinical course of meningitis was aggravated in mice lacking TLR2. In this pneumococcal infection model, the worsening of disease was accompanied by a higher bacterial load in the brain and a more pronounced increase in both cerebral TNF-α activity and blood–brain barrier disruption (Echchannaoui et al. 2002). The observations that TLR2 deficiency led to a moderate increase in disease severity, cerebral bacterial titers and blood–brain barrier compromise were confirmed by our experiments in which pneumococcal meningitis was induced by direct inoculation of the pathogen into the CSF compartment (Koedel et al. 2003). Supplemental analyses of the inflammatory host response in our meningitis model revealed that, in early meningitis (4 h after pneumococcal infection), brain expression of TNF-α and the CXC chemokine CXCL-2 (but not of IL-1β, inducible nitric oxide synthase, and complement C3) was significantly lower in TLR2-deficient mice than in wild-type mice. However, in later stages of the disease, the brain cytokine and chemokine expression levels did not differ between the two mouse strains. Accordingly, CSF leukocyte counts were similar in wild-type and TLR2-deficient mice (Koedel et al. 2003). These observations hinted at higher pneumococcal toxin concentrations (e.g., pneumolysin and hydrogen peroxide) (Braun et al. 2002) as a major factor in deterioration to more clinical symptoms in TLR2-deficient mice. Moreover, the increased release of pneumococcal toxins due to the increased bacterial burden in TLR2-deficient mice was suggested to compensate for the anti-inflammatory properties exerted by TLR2 deficiency. Finally, since the immune milieu in the CSF compartment is different to that of the brain (as depicted above), the site of pneumococcal inoculation may account for the differences observed in brain TNF-α expression between our experiments and those by Echchannaoui et al. (2002). Altogether, these initial studies argued for an important role of TLR2 in host defense against pneumococcal infection in the CSF, but also for the involvement of further PRRs in mounting the host immune response in pneumococcal meningitis. In order to get a better insight into the role of TLR signaling in this disease, we then used gene-targeted mice lacking functional MyD88, which is used by all TLRs apart from TLR3. Compared to TLR2-deficient mice, the phenotype of MyD88-deficient mice in our meningitis model was much more impressive (Koedel et al. 2004). For example, while none of the wild-type mice died during the first 24 h after pneumococcal infection, the mortality rate of infected MyD88-deficient mice was approximately 45%. The more adverse outcome observed in MyD88-deficient mice was paralleled by a dramatically impaired host defense in both the CSF and the periphery, as evidenced by substantially higher bacterial loads in the brain, blood and lung. The higher susceptibility to intracisternal pneumococcal infection was due to a defective host immune response inside the CSF. Compared to infected wild-type mice, MyD88-deficient mice showed an 80% reduction in CSF leukocyte counts, which was associated with a near-complete abrogation of the brain expression of proinflammatory cytokines and chemokines (Fig. 3). Since the substantial attenuation of meningeal inflammation resulted in a marked reduction of meningitis-induced intracranial complication (such as blood–brain barrier compromise), it is most likely that the worsening of pneumococcal disease in MyD88-deficient mice is due to the severe bacteremia and

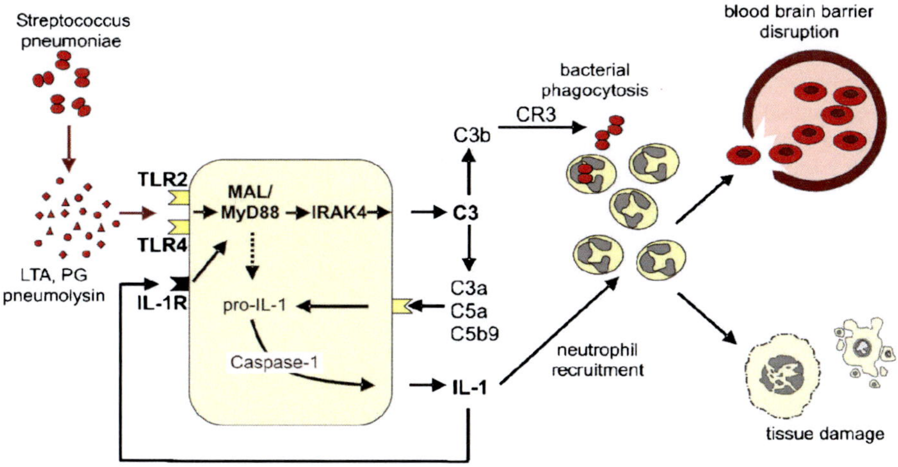

Fig. 3 Schematic diagram of the main pathogenic steps of pneumococcal meningitis. Inside the CSF compartment, the presence of *S. pneumoniae* seems to be recognized initially by TLR2 and TLR4 (as well as by as-yet unidentified pattern recognition receptors). TLR signaling via MyD88/Mal results in the induction of complement factor 3 (C3) expression, which is absent from normal CSF. After its activation/cleavage, C3 contributes to the killing of pneumococci, but also promotes the immune response, presumably via upregulation of the expression of IL-1 family cytokines, which are essential for the recruitment of neutrophils to the site of infection. An autocrine (and possibly paracrine) positive feedback loop via IL-1 and MyD88 contributes to the exaggeration of the host inflammatory reaction, which is a key mediator of meningitis-associated blood–brain barrier disruption and tissue injury

the aggravation of internal complications including sepsis and pneumonia (Koedel et al. 2004). These data strongly indicated a fundamental role of MyD88 in the immune reaction to pneumococcal infection of the CSF compartment. The fact that MyD88 is an adapter molecule common to signaling pathways of IL-1Rs and all TLRs activated in vitro by *S. pneumoniae* suggested that the phenotype of MyD88-deficient mice can be linked to defects in either the IL-1 pathway or TLR-mediated signaling, or in both. By utilizing mice with single or combined deficiencies of TLR2, TLR4, and TLR9, all three of which were implicated in the sensing of *S. pneumoniae* in vitro, we were able to demonstrate that the concerted action of both TLR2 and TLR4 plays a crucial role in host defense and immune activation in pneumococcal meningitis (Klein et al. 2008). TLR2-TLR4 double-deficient mice had a 50% reduction in CSF pleocytosis and a selective inhibition of cytokine production. Moreover, similar to the TLR2-deficient state, the single deficiency in TLR4 was found to have no significant impact on either bacterial outgrowth in the CSF or on meningeal inflammation. However, contrary to our expectations deduced from the vitro findings, our in vivo analyses did not suggest TLR9 as an additional sensor for pneumococcal infection of the CSF, since additional TLR9 deficiency did not result in further attenuation of the inflammatory reaction observed in TLR2-TLR4 double-deficient mice (Klein et al. 2008). The observation that TLR2-TLR4 double-deficient mice were less severely impaired in their immune response than MyD88-deficient mice suggests that the MyD88 phenotype may be—at least partly—due to the blocking of secondary autocrine effects of IL-1 family cytokines. This concept is supported by recent studies in mice with deficiencies of individual genes of the IL-1/IL-18 pathway. Recently, we demonstrated that targeted disruption of the gene for the cysteine protease caspase-1, which is crucial for the generation of both active IL-1 and IL-18 significantly diminished the inflammatory host response to pneumococci in the CSF compartment (Koedel et al. 2002b).

Using a mouse model of hematogeneous meningitis, Zwijnenburg et al. (2003b,a) provided further evidence for the crucial role of both IL-1 and IL-18 pathways in the immune regulation in pneumococcal meningitis. In IL-1R-deficient mice, pneumococcal meningitis was associated with less severe leukocyte infiltration and with lower brain levels of cytokines and chemokines (Zwijnenburg et al. 2003b). IL-18-deficient mice were also reported to show a suppressed inflammatory response, as evidenced by a less profound inflammatory infiltrate as well as lower brain cytokine and chemokine levels (Zwijnenburg et al. 2003a). Thus, an interrupted autocrine feedback loop via IL-1 receptors is likely to contribute to the phenotype of MyD88-deficient mice. The fact that MyD88 deficiency resulted in strong but not complete inhibition of the host immune response furthermore indicated the presence of additional PRRs in the recognition of *S. pneumoniae*. Cell culture experiments indicated that internalized *S. pneumoniae* can be sensed by NOD2 (Opitz et al. 2004). However, mice lacking RIP2, which is crucial for NOD1- and NOD2-dependent immune activation (Park et al. 2007), did not show any alterations in the host immune response to *S. pneumoniae* in our meningitis model (unpublished data). This finding is strengthened by a recent study which monitored the trafficking of pneumococcal cell walls (PCW) in the host as well as its consequences on the host after intravascular injection (Fillon et al. 2006). Intravascular PCW was found to bind to endothelial cells and cause rapid lethality in both wild-type and NOD2-deficient mice to a similar degree. Since PCW was also observed to be internalized into host cells, these data argue against an important role of NOD2 in the sensing of pneumococcal fragments in vivo.

Collectively, experiments using mouse meningitis models supported the idea that TLRs are crucial for the detection of pathogens in the CSF compartment. Contrary to the in vitro observations, only TLR2 and TLR4, not TLR9, act as sensors of *S. pneumoniae* infection of the CSF. Engagement of these TLRs by pneumococcal ligands leads to MyD88-dependent production of proinflammatory cytokines of the IL-1 family. Secretion of IL-1 family cytokines forms a positive feedback loop that boosts MyD88-dependent production of inflammatory mediators. The impressive phenotype of MyD88-deficient mice illustrates the extraordinary importance of TLR-IL-1R-MyD88 signaling in regulating the host immune response to pneumococcal infection of the CSF compartment.

Based upon the experimental evidence described in this article, molecular genetic studies have been performed in humans to assess whether genes of the TLR pathway influence immunity to pneumococcal infection. A listing of all TLR-related gene polymorphisms and studies investigating associations with susceptibility to pneumococcal diseases is beyond the scope of this chapter, but two examples are mentioned here. First, children with an inherited deficiency of the IL-1 receptor kinase (IRAK)-4, which is essential for NF-κB activation in TLR- and IL-1R signaling pathways, are highly susceptible to severe (and often recurrent) invasive pneumococcal disease (Ku et al. 2007). Secondly, heterozygous carriage of a leucine substitution at serine 180 of Mal, another key molecule in the TLR/IL-1R signal transduction cascade, was reported to roughly halve the risk for invasive pneumococcal disease (Khor et al. 2007). The authors speculated that heterozygosity of Mal may confer protective immunity, characterized by an intermediate TLR/IL-1R activation state and a well-balanced inflammatory response, whereas homozygosity of Mal (or IRAK-4) may result in impaired host-defense mechanisms and overwhelming infection. Altogether, these studies suggest that TLR/IL-1R pathways are vital for immunity to *S. pneumoniae*.

6 Conclusions

Over the past 20 years it has become clear that intracranial complications (including brain edema formation and alterations in cerebral blood flow) are major determinants of an unfavorable outcome in bacterial meningitis, and are largely caused by the host immune response (Koedel et al. 2002a; Weber and Tuomanen 2007). Recent studies showed that direct injection of synthetic TLR2 or TLR9 agonists can induce inflammatory and associated neuropathological changes typically associated with bacterial meningitis (Deng et al. 2001; Hoffmann et al. 2007). Moreover, targeted disruption of both TLR2 and TLR4 led to a substantial reduction of meningitis-induced intracranial complications and tissue damage in a murine model of pneumococcal meningitis (Klein et al. 2008). The neuroprotective effects of TLR deficiency, together with the meningitis-inducing potency of TLR agonists, strongly suggest that targeted interference with TLR signaling is a promising strategy for dampening meningeal inflammation and thus improving the outcome of the disease. However, the road toward the ultimate goal of helping patients with meningitis by using TLR antagonists is still long. First of all, in vivo studies are absolutely warranted to clarify the mechanisms of immune activation in pathogen recognition in bacterial meningitis due to other pathogens besides *S. pneumoniae*. While in vitro data indicate a crucial role of TLRs in the detection of *N. meningitidis* infection, their contribution to immune activation upon infection with either *S. agalactiae* or *L. monocytogenes* in vivo is largely unclear and needs to be addressed in future studies using mutant mice with single or combined deficiencies in genes of the TLR signaling cascade. Moreover, based on studies performed using pneumococcal meningitis models, TLRs are likely to be key triggers of the immune response to the pathogen, but TLR-independent, IL-1R-dependent MyD88-signaling pathways seem to be crucial for the aggravation and, quite possibly, the perpetuation of meningeal inflammation. Therefore, it must be evaluated whether pharmacological interference with TLR signaling is still effective at reducing inflammation and tissue injury once the inflammatory reaction has been initiated. In this context, it is of special interest to assess the efficacy of TLR antagonists (e.g., neutralizing antibodies) as adjuvant therapy (given either simultaneously with or at defined time points after the start of antibiotic therapy) in experimental bacterial meningitis. Finally, targeted disruptions of either MyD88 or TLR2/4 were found to result in the worsening of disease in a mouse model of pneumococcal meningitis. Disease deterioration was associated with severe bacteremia, presumably leading to an aggravation of internal complications such as septic shock and pneumonia. These data make it conceivable that the benefit to the brain of TLR antagonism may be overshadowed by the risk of impaired bacterial eradication and uncontrolled infection of the host. Therefore, studies evaluating whether adjuvant therapy resulting from TLR antagonists impairs the penetration and action of antibiotics in the CSF are strongly warranted. Deciphering the exact role of TLRs (and other PRRs) in bacterial eradication, immune activation and immune resolution in the CSF compartment is a challenge for future meningitis research.

Acknowledgments I warmly thank Prof. C. Kirschning and Prof. H. Wagner (Institute of Medical Microbiology, Immunology, and Hygiene, Technical University, Munich, Germany) as well as Prof. J. Heesemann and Prof. S. Hammerschmidt (Max-von-Pettenkofer Institute for Hygiene and Microbiology, Ludwig-Maximilians University, Munich, Germany) for their excellent cooperation over many years. I further thank Prof. H.W. Pfister, Dr. R. Paul and Dr. M. Klein (Department of Neurology, Ludwig-Maximilians University, Munich, Germany) for helpful discussions and Ms. K. Ogston for copyediting the manuscript. Last but not least, the financial support of the Deutsche Forschungs-Gemeinschaft, the Else Kroener-Fresenius Stiftung, and the Meningitis Research Foundation is greatly appreciated.

References

Akira S, Uematsu S, Takeuchi O (2006) Pathogen recognition and innate immunity. Cell 124:783–801

Barton GM, Medzhitov R (2003) Toll-like receptor signaling pathways. Science 300:1524–1525

Benton KA, Paton JC, Briles DE (1997) Differences in virulence for mice among *Streptococcus pneumoniae* strains of capsular types 2, 3, 4, 5, and 6 are not attributable to differences in pneumolysin production. Infect Immun 65:1237–1244

Bercion R, Bobossi-Serengbe G, Gody JC, Beyam EN, Manirakiza A, Le Faou A (2008) Acute bacterial meningitis at the "Complexe Pediatrique" of Bangui, Central African republic. J Trop Pediatr 154:125–128

Beutler B, Jiang Z, Georgel P, Crozat K, Croker B, Rutschmann S, Du X, Hoebe K (2006) Genetic analysis of host resistance: Toll-like receptor signaling and immunity at large. Annu Rev Immunol 24:353–389

Beutler B, Rietschel ET (2003) Innate immune sensing and its roots: the story of endotoxin. Nat Rev Immunol 3:169–176

Bingen E, Levy C, de la Rocque F, Boucherat M, Varon E, Alonso JM, Dabernat H, Reinert P, Aujard Y, Cohen R (2005) Bacterial meningitis in children: a French prospective study. Clin Infect Dis 41:1059–1063

Brandtzaeg P, Ovsteboo R, Kierulf P (1992) Compartmentalization of lipopolysaccharide production correlates with clinical presentation in meningococcal disease. J Infect Dis 166:650–652

Braun JS, Sublett JE, Freyer D, Mitchell TJ, Cleveland JL, Tuomanen EI, Weber JR (2002) Pneumococcal pneumolysin and H_2O_2 mediate brain cell apoptosis during meningitis. J Clin Invest 109:19–27

Bsibsi M, Ravid R, Gveric D, van Noort JM (2002) Broad expression of Toll-like receptors in the human central nervous system. J Neuropathol Exp Neurol 61:1013–1021

De Luigi A, Terreni L, Sironi M, De Simoni MG (1998) The sympathetic nervous system tonically inhibits peripheral interleukin-1beta and interleukin-6 induction by central lipopolysaccharide. Neuroscience 83:1245–1250

Deng GM, Liu ZQ, Tarkowski A (2001) Intracisternally localized bacterial DNA containing CPG motifs induces meningitis. J Immunol 167:4616–4626

Dery M, Hasbun R (2007) Changing epidemiology of bacterial meningitis. Curr Infect Dis Rep 9:301–307

Doganay M (2003) Listeriosis: clinical presentation. FEMS Immunol Med Microbiol 35:173–175

Draper DW, Bethea HN, He YW (2006) Toll-like receptor 2-dependent and -independent activation of macrophages by group B streptococci. Immunol Lett 102:202–214

Dujardin BC, Driedijk PC, Roijers AF, Out TA (1985) The determination of the complement components C1q, C4 and C3 in serum and cerebrospinal fluid by radioimmunoassay. J Immunol Methods 80:227–237

Echchannaoui H, Frei K, Schnell C, Leib SL, Zimmerli W, Landmann R (2002) Toll-like receptor 2-deficient mice are highly susceptible to *Streptococcus pneumoniae* meningitis because of reduced bacterial clearing and enhanced inflammation. J Infect Dis 186:798–806

Fillon S, Soulis K, Rajasekaran S, Beneedict-Hamilton H, Radin JN, Orihuela CJ, El Kasmi KC, Murti G, Kaushal D, Gaber MW, Weber JR, Murray PJ, Tuomanen EI (2006) Platelet-activating factor receptor and innate immunity: uptake of gram-positive bacterial cell wall into host cells and cell-specific pathophysiology. J Immunol 177:6182–6191

Fisette PL, Ram S, Andersen JM, Guo W, Ingalls RR (2003) The Lip lipoprotein from *Neisseria gonorrhoeae* stimulates cytokine release and NF-kappaB activation in epithelial cells in a Toll-like receptor 2-dependent manner. J Biol Chem 278:46252–46260

Flo TH, Halaas O, Lien E, Ryan L, Teti G, Golenbock DT, Sundan A, Espevik T (2000) Human toll-like receptor 2 mediates monocyte activation by *Listeria monocytogenes*, but not by group B streptococci or lipopolysaccharide. J Immunol 164:2064–2069

Franchi L, Kanneganti TD, Dubyak GR, Nunez G (2007) Differential requirement of P2X7 receptor and intracellular K+ for caspase-1 activation induced by intracellular and extracellular bacteria. J Biol Chem 282:18810–18818

Galea I, Palin K, Newman TA, van RN, Perry VH, Boche D (2005) Mannose receptor expression specifically reveals perivascular macrophages in normal, injured, and diseased mouse brain. Glia 49:375–384

Gerber J, Raivich G, Wellmer A, Noeske C, Kunst T, Werner A, Bruck W, Nau R (2001) A mouse model of *Streptococcus pneumoniae* meningitis mimicking several features of human disease. Acta Neuropathol (Berl) 101:499–508

Girardin SE, Boneca IG, Carneiro LA, Antignac A, Jehanno M, Viala J, Tedin K, Taha MK, Labigne A, Zathringer U, Coyle AJ, DiStefano PS, Bertin J, Sansonetti PJ, Philpott DJ (2003a) Nod1 detects a unique muropeptide from gram-negative bacterial peptidoglycan. Science 300:1584–1587

Girardin SE, Boneca IG, Viala J, Chamaillard M, Labigne A, Thomas G, Philpott DJ, Sansonetti PJ (2003b) Nod2 is a general sensor of peptidoglycan through muramyl dipeptide (MDP) detection. J Biol Chem 278:8869–8872

Guillemin GJ, Brew BJ (2004) Microglia, macrophages, perivascular macrophages, and pericytes: a review of function and identification. J Leukoc Biol 75:388–397

Han SH, Kim JH, Martin M, Michalek SM, Nahm MH (2003) Pneumococcal lipoteichoic acid (LTA) is not as potent as staphylococcal LTA in stimulating Toll-like receptor 2. Infect Immun 71:5541–5548

Hanisch UK, Prinz M, Angstwurm K, Hausler KG, Kann O, Kettenmann H, Weber JR (2001) The protein tyrosine kinase inhibitor AG126 prevents the massive microglial cytokine induction by pneumococcal cell walls. Eur J Immunol 31:2104–2115

Hasegawa M, Yang K, Hashimoto M, Park JH, Kim YG, Fujimoto Y, Nunez G, Fukase K, Inohara N (2006) Differential release and distribution of Nod1 and Nod2 immunostimulatory molecules among bacterial species and environments. J Biol Chem 281:29054–29063

Hayashi F, Smith KD, Ozinsky A, Hawn TR, Yi EC, Goodlett DR, Eng JK, Akira S, Underhill DM, Aderem A (2001) The innate immune response to bacterial flagellin is mediated by Toll-like receptor 5. Nature 410:1099–1103

Heath PT, Nik Yusoff NK, Baker CJ (2003) Neonatal meningitis. Arch Dis Child Fetal Neonatal Ed 88:F173–F178

Heffron R (1979) Pneumonia with special reference to pneumococcus lobar pneumonia. Harvard University Press, Cambridge

Henneke P, Takeuchi O, Malley R, Lien E, Ingalls RR, Freeman MW, Mayadas T, Nizet V, Akira S, Kasper DL, Golenbock DT (2002) Cellular activation, phagocytosis, and bactericidal activity against group B streptococcus involve parallel myeloid differentiation factor 88-dependent and independent signaling pathways. J Immunol 169:3970–3977

Henneke P, Takeuchi O, van Strijp JA, Guttormsen HK, Smith JA, Schromm AB, Espevik TA, Akira S, Nizet V, Kasper DL, Golenbock DT (2001) Novel engagement of CD14 and multiple Toll-like receptors by group B streptococci. J Immunol 167:7069–7076

Heumann D, Barras C, Severin A, Glauser MP, Tomasz A (1994) Gram-positive cell walls stimulate synthesis of tumor necrosis factor alpha and interleukin-6 by human monocytes. Infect Immun 62:2715–2721

Hoffmann O, Braun JS, Becker D, Halle A, Freyer D, Dagand E, Lehnardt S, Weber JR (2007) TLR2 mediates neuroinflammation and neuronal damage. J Immunol 178:6476–6481

Janot L, Secher T, Torres D, Maillet I, Pfeilschifter J, Quesniaux VF, Landmann R, Ryffel B, Erard F (2008) CD14 works with Toll-Like Receptor 2 to contribute to recognition and control of *Listeria monocytogenes* infection. J Infect Dis 198:115–124

Jefferies JM, Johnston CH, Kirkham LA, Cowan GJ, Ross KS, Smith A, Clarke SC, Brueggemann AB, George RC, Pichon B, Pluschke G, Pfluger V, Mitchell TJ (2007) Presence of nonhemolytic pneumolysin in serotypes of *Streptococcus pneumoniae* associated with disease outbreaks. J Infect Dis 196:936–944

Johnston M, Zakharov A, Papaiconomou C, Salmasi G, Armstrong D (2004) Evidence of connections between cerebrospinal fluid and nasal lymphatic vessels in humans, non-human primates and other mammalian species. Cerebrospinal Fluid Res 1:2

Jounblat R, Kadioglu A, Mitchell TJ, Andrew PW (2003) Pneumococcal behavior and host responses during bronchopneumonia are affected differently by the cytolytic and complement-activating activities of pneumolysin. Infect Immun 71:1813–1819

Kanneganti TD, Lamkanfi M, Nunez G (2007) Intracellular NOD-like receptors in host defense and disease. Immunity 27:549–559

Kastenbauer S, Pfister HW (2003) Pneumococcal meningitis in adults: spectrum of complications and prognostic factors in a series of 87 cases. Brain 126:1015–1025

Khor CC, Chapman SJ, Vannberg FO, Dunne A, Murphy C, Ling EY, Frodsham AJ, Walley AJ, Kyrieleis O, Khan A, Aucan C, Segal S, Moore CE, Knox K, Campbell SJ, Lienhardt C, Scott A, Aaby P, Sow OY, Grignani RT, Sillah J, Sirugo G, Peshu N, Williams TN, Maitland K, Davies RJ, Kwiatkowski DP, Day NP, Yala D, Crook DW, Marsh K, Berkley JA, O'Neill LA, Hill AV (2007) A Mal functional variant is associated with protection against invasive pneumococcal disease, bacteremia, malaria and tuberculosis. Nat Genet 39:523–528

Kim YG, Park JH, Shaw MH, Franchi L, Inohara N, Nunez G (2008) The cytosolic sensors Nod1 and Nod2 are critical for bacterial recognition and host defense after exposure to Toll-like receptor ligands. Immunity 28:246–257

Klein M, Angele B, Pfister HW, Wagner H, Koedel U, Kirschning CJ (2008) Detection of pneumococcal infection of the central nervous system depends upon TLR2 and TLR4. J Infect Dis 198:1028–1036

Koedel U, Angele B, Rupprecht T, Wagner H, Roggenkamp A, Pfister HW, Kirschning CJ (2003) Toll-Like receptor 2 participates in mediation of immune response in experimental pneumococcal meningitis. J Immunol 170:438–444

Koedel U, Rupprecht T, Angele B, Heesemann J, Wagner H, Pfister HW, Kirschning CJ (2004) MyD88 is required for mounting a robust host immune response to *Streptococcus pneumoniae* in the CNS. Brain 127:1437–1445

Koedel U, Scheld WM, Pfister HW (2002a) Pathogenesis and pathophysiology of pneumococcal meningitis. Lancet Infect Dis 2:721–736

Koedel U, Winkler F, Angele B, Fontana A, Flavell RA, Pfister HW (2002b) Role of Caspase-1 in experimental pneumococcal meningitis: evidence from pharmacologic caspase inhibition and caspase-1-deficient mice. Ann Neurol 51:319–329

Ku CL, von BH, Picard C, Zhang SY, Chang HH, Yang K, Chrabieh M, Issekutz AC, Cunningham CK, Gallin J, Holland SM, Roifman C, Ehl S, Smart J, Tang M, Barrat FJ, Levy O, McDonald D, Day-Good NK, Miller R, Takada H, Hara T, Al-Hajjar S, Al-Ghonaium A, Speert D, Sanlaville D, Li X, Geissmann F, Vivier E, Marodi L, Garty BZ, Chapel H, Rodriguez-Gallego C, Bossuyt X, Abel L, Puel A, Casanova JL (2007) Selective predisposition to bacterial infections in IRAK-4-deficient children: IRAK-4-dependent TLRs are otherwise redundant in protective immunity. J Exp Med 204:2407–2422

Laflamme N, Echchannaoui H, Landmann R, Rivest S (2003) Cooperation between toll-like receptor 2 and 4 in the brain of mice challenged with cell wall components derived from gram-negative and gram-positive bacteria. Eur J Immunol 33:1127–1138

Laflamme N, Rivest S (2001) Toll-like receptor 4: the missing link of the cerebral innate immune response triggered by circulating gram-negative bacterial cell wall components. FASEB J 15:155–163

Laflamme N, Soucy G, Rivest S (2001) Circulating cell wall components derived from gram-negative, not gram-positive, bacteria cause a profound induction of the gene-encoding Toll-like receptor 2 in the CNS. J Neurochem 79:648–657

Lata S, Raghava GP (2008) PRRDB: a comprehensive database of pattern-recognition receptors and their ligands. BMC Genomics 9:180

Leber JH, Crimmins GT, Raghavan S, Meyer-Morse NP, Cox JS, Portnoy DA (2008) Distinct TLR- and NLR-mediated transcriptional responses to an intracellular pathogen. PLoS Pathog 4(1):e6

Lee KS, Scanga CA, Bachelder EM, Chen Q, Snapper CM (2007) TLR2 synergizes with both TLR4 and TLR9 for induction of the MyD88-dependent splenic cytokine and chemokine response to *Streptococcus pneumoniae*. Cell Immunol 245:103–110

Lemaitre B, Nicolas E, Michaut L, Reichhart JM, Hoffmann JA (1996) The dorsoventral regulatory gene cassette spatzle/Toll/cactus controls the potent antifungal response in *Drosophila* adults. Cell 86:973–983

Letiembre M, Echchannaoui H, Bachmann P, Ferracin F, Nieto C, Espinosa M, Landmann R (2005) Toll-like receptor 2 deficiency delays pneumococcal phagocytosis and impairs oxidative killing by granulocytes. Infect Immun 73:8397–8401

Letiembre M, Hao W, Liu Y, Walter S, Mihaljevic I, Rivest S, Hartmann T, Fassbender K (2007) Innate immune receptor expression in normal brain aging. Neuroscience 146:248–254

Magnusson M, Tobes R, Sancho J, Pareja E (2007) Cutting edge: natural DNA repetitive extragenic sequences from gram-negative pathogens strongly stimulate TLR9. J Immunol 179:31–35

Makwana N, Riordan FA (2007) Bacterial meningitis: the impact of vaccination. CNS Drugs 21:355–366

Malley R, Henneke P, Morse SC, Cieslewicz MJ, Lipsitch M, Thompson CM, Kurt-Jones E, Paton JC, Wessels MR, Golenbock DT (2003) Recognition of pneumolysin by Toll-like receptor 4 confers resistance to pneumococcal infection. Proc Natl Acad Sci USA 100:1966–1971

Mariathasan S, Weiss DS, Newton K, McBride J, O'Rourke K, Roose-Girma M, Lee WP, Weinrauch Y, Monack DM, Dixit VM (2006) Cryopyrin activates the inflammasome in response to toxins and ATP. Nature 440:228–232

Massari P, Henneke P, Ho Y, Latz E, Golenbock DT, Wetzler LM (2002) Cutting edge: immune stimulation by neisserial porins is toll-like receptor 2 and MyD88 dependent. J Immunol 168:1533–1537

Massari P, Visintin A, Gunawardana J, Halmen KA, King CA, Golenbock DT, Wetzler LM (2006) Meningococcal porin PorB binds to TLR2 and requires TLR1 for signaling. J Immunol 176:2373–2380

Matyszak MK, Perry VH (1996) A comparison of leucocyte responses to heat-killed bacillus Calmette–Guérin in different CNS compartments. Neuropathol Appl Neurobiol 22:44–53

McGreal EP, Miller JL, Gordon S (2005) Ligand recognition by antigen-presenting cell C-type lectin receptors. Curr Opin Immunol 17:18–24

Medzhitov R, Janeway CAJ (2002) Decoding the patterns of self and nonself by the innate immune system. Science 296:298–300

Medzhitov R, Preston-Hurlburt P, Janeway CAJ (1997) A human homologue of the *Drosophila* Toll protein signals activation of adaptive immunity. Nature 388:394–397

Mogensen TH, Paludan SR, Kilian M, Ostergaard L (2006) Live *Streptococcus pneumoniae*, *Haemophilus influenzae*, and *Neisseria meningitidis* activate the inflammatory response through Toll-like receptors 2, 4, and 9 in species-specific patterns. J Leukoc Biol 80:267–277

Mukhopadhyay S, Gordon S (2004) The role of scavenger receptors in pathogen recognition and innate immunity. Immunobiology 209:39–49

Naito M, Kodama T, Matsumoto A, Doi T, Takahashi K (1991) Tissue distribution, intracellular localization, and in vitro expression of bovine macrophage scavenger receptors. Am J Pathol 139:1411–1423

Niederkorn JY (2006) See no evil, hear no evil, do no evil: the lessons of immune privilege. Nat Immunol 7:354–359

Ogura Y, Sutterwala FS, Flavell RA (2006) The inflammasome: first line of the immune response to cell stress. Cell 126:659–662

Opitz B, Puschel A, Schmeck B, Hocke AC, Rosseau S, Hammerschmidt S, Schumann RR, Suttorp N, Hippenstiel S (2004) Nucleotide-binding oligomerization domain proteins are innate immune receptors for internalized *Streptococcus pneumoniae*. J Biol Chem 279:36426–36432

Pachter JS, De Vries HE, Fabry Z (2003) The blood–brain barrier and its role in immune privilege in the central nervous system. J Neuropathol Exp Neurol 62:593–604

Park JH, Kim YG, McDonald C, Kanneganti TD, Hasegawa M, Body-Malapel M, Inohara N, Nunez G (2007) RICK/RIP2 mediates innate immune responses induced through Nod1 and Nod2 but not TLRs. J Immunol 178:2380–2386

Pashenkov M, Link H (2002) Dendritic cells and immune responses in the central nervous system. Trends Immunol 23:69–70

Poltorak A, He X, Smirnova I, Liu MY, Huffel CV, Du X, Birdwell D, Alejos E, Silva M, Galanos C, Freudenberg M, Ricciardi CP, Layton B, Beutler B (1998) Defective LPS signaling in C3H/HeJ and C57BL/10ScCr mice: mutations in *Tlr4* gene. Science 282:2085–2088

Pridmore AC, Jarvis GA, John CM, Jack DL, Dower SK, Read RC (2003) Activation of toll-like receptor 2 (TLR2) and TLR4/MD2 by *Neisseria* is independent of capsule and lipooligosaccharide (LOS) sialylation but varies widely among LOS from different strains. Infect Immun 71:3901–3908

Pridmore AC, Wyllie DH, Abdillahi F, Steeghs L, van der Ley P, Dower SK, Read RC (2001) A lipopolysaccharide-deficient mutant of *Neisseria meningitidis* elicits attenuated cytokine release by human macrophages and signals via toll-like receptor (TLR) 2 but not via TLR4/MD2. J Infect Dis 183:89–96

Riesenfeld-Orn I, Wolpe S, Garcia-Bustos JF, Hoffmann MK, Tuomanen E (1989) Production of interleukin-1 but not tumor necrosis factor by human monocytes stimulated with pneumococcal cell surface components. Infect Immun 57:1890–1893

Rijneveld AW, Florquin S, Branger J, Speelman P, van Deventer SJ, Van der Poll T (2001) TNF-alpha compensates for the impaired host defense of IL-1 type I receptor-deficient mice during pneumococcal pneumonia. J Immunol 167:5240–5246

Robbins JB, Towne DW, Gotschlich EC, Schneerson R (1997) "Love's labours lost": failure to implement mass vaccination against group A meningococcal meningitis in sub-Saharan Africa. Lancet 350:880–882

Rodriguez-Martinez S, Cancino-Diaz ME, Jimenez-Zamudio L, Garcia-Latorre E, Cancino-Diaz JC (2005) TLRs and NODs mRNA expression pattern in healthy mouse eye. Br J Ophthalmol 89:904–910

Roozendaal R, Carroll MC (2006) Emerging patterns in complement-mediated pathogen recognition. Cell 125:29–32

Rossi IA, Zuber PL, Dumolard L, Walker DG, Watt J (2007) Introduction of Hib vaccine into national immunization programmes: a descriptive analysis of global trends. Vaccine 25:7075–7080

Schmeck B, Huber S, Moog K, Zahlten J, Hocke AC, Opitz B, Hammerschmidt S, Mitchell TJ, Kracht M, Rosseau S, Suttorp N, Hippenstiel S (2006) Pneumococci induced TLR- and Rac1-dependent NF-kappaB-recruitment to the IL-8 promoter in lung epithelial cells. Am J Physiol Lung Cell Mol Physiol 290:L730–L737

Schneider O, Michel U, Zysk G, Dubuis O, Nau R (1999) Clinical outcome in pneumococcal meningitis correlates with CSF lipoteichoic acid concentrations. Neurology 53:1584–1587

Schroder NW, Morath S, Alexander C, Hamann L, Hartung T, Zahringer U, Gobel UB, Weber JR, Schumann RR (2003) Lipoteichoic acid (LTA) of *Streptococcus pneumoniae* and *Staphylococcus aureus* activates immune cells via Toll-like receptor (TLR)-2, lipopolysaccharide-binding protein (LBP), and CD14, whereas TLR-4 and MD-2 are not involved. J Biol Chem 278:15587–15594

Schuchat A, Robinson K, Wenger JD, Harrison LH, Farley M, Reingold AL, Lefkowitz L, Perkins BA (1997) Bacterial meningitis in the United States in 1995. Active Surveillance Team. N Engl J Med 337:970–976

Schut ES, de GJ, van de BD (2008) Community-acquired bacterial meningitis in adults. Pract Neurol 8:8–23

Schwartz M, Butovsky O, Bruck W, Hanisch UK (2006) Microglial phenotype: is the commitment reversible? Trends Neurosci 29:68–74

Seki E, Tsutsui H, Tsuji NM, Hayashi N, Adachi K, Nakano H, Futatsugi-Yumikura S, Takeuchi O, Hoshino K, Akira S, Fujimoto J, Nakanishi K (2002) Critical roles of myeloid differentiation factor 88-dependent proinflammatory cytokine release in early phase clearance of *Listeria monocytogenes* in mice. J Immunol 169:3863–3868

Singhrao SK, Neal JW, Rushmere NK, Morgan BP, Gasque P (1999) Differential expression of individual complement regulators in the brain and choroid plexus. Lab Invest 79:1247–1259

Stahel PF, Nadal D, Pfister HW, Paradisis PM, Barnum SR (1997) Complement C3 and factor B cerebrospinal fluid concentrations in bacterial and aseptic meningitis. Lancet 349:1886–1887

Steinman RM (2007) Lasker Basic Medical Research Award. Dendritic cells: versatile controllers of the immune system. Nat Med 13:1155–1159

Swartz MN (2004) Bacterial meningitis: a view of the past 90 years. N Engl J Med 351:1826–1828

Syrogiannopoulos GA, Hansen EJ, Erwin AL, Munford RS, Rutledge J, Reisch JS, McCracken GHJ (1988) *Haemophilus influenzae* type b lipooligosaccharide induces meningeal inflammation. J Infect Dis 157:237–244

Takaoka A, Wang Z, Choi MK, Yanai H, Negishi H, Ban T, Lu Y, Miyagishi M, Kodama T, Honda K, Ohba Y, Taniguchi T (2007) DAI (DLM-1/ZBP1) is a cytosolic DNA sensor and an activator of innate immune response. Nature 448:501–505

Takeuchi O, Akira S (2002) MyD88 as a bottle neck in Toll/IL-1 signaling. Curr Top Microbiol Immunol 270:155–167

Thwaites G, Chau TT, Mai NT, Drobniewski F, McAdam K, Farrar J (2000) Tuberculous meningitis. J Neurol Neurosurg Psychiatry 68:289–299

Tsai CJ, Griffin MR, Nuorti JP, Grijalva CG (2008) Changing epidemiology of pneumococcal meningitis after the introduction of pneumococcal conjugate vaccine in the United States. Clin Infect Dis 46:1664–1672

Tuomanen E, Hengstler B, Zak O, Tomasz A (1986) Induction of meningeal inflammation by diverse bacterial cell walls. Eur J Clin Microbiol 5:682–684

Tuomanen E, Liu H, Hengstler B, Zak O, Tomasz A (1985a) The induction of meningeal inflammation by components of the pneumococcal cell wall. J Infect Dis 151:859–868

Tuomanen E, Tomasz A, Hengstler B, Zak O (1985b) The relative role of bacterial cell wall and capsule in the induction of inflammation in pneumococcal meningitis. J Infect Dis 151:535–540

Tuomanen EI, Saukkonen K, Sande S, Cioffe C, Wright SD (1989) Reduction of inflammation, tissue damage, and mortality in bacterial meningitis in rabbits treated with monoclonal antibodies against adhesion-promoting receptors of leukocytes. J Exp Med 170:959–969

Van de Beek D, De Gans J, Spanjaard L, Weisfelt M, Reitsma JB, Vermeulen M (2004) Clinical features and prognostic factors in adults with bacterial meningitis. N Engl J Med 351:1849–1859

Waage A, Halstensen A, Shalaby R, Brandtzaeg P, Kierulf P, Espevik T (1989) Local production of tumor necrosis factor alpha, interleukin 1, and interleukin 6 in meningococcal meningitis. Relation to the inflammatory response. J Exp Med 170:1859–1867

Wang Z, Choi MK, Ban T, Yanai H, Negishi H, Lu Y, Tamura T, Takaoka A, Nishikura K, Taniguchi T (2008) Regulation of innate immune responses by DAI (DLM-1/ZBP1) and other DNA-sensing molecules. Proc Natl Acad Sci USA 105:5477–5482

Warren SE, Mao DP, Rodriguez AE, Miao EA, Aderem A (2008) Multiple Nod-like receptors activate caspase 1 during *Listeria monocytogenes* infection. J Immunol 180:7558–7564

Weber JR, Tuomanen EI (2007) Cellular damage in bacterial meningitis: an interplay of bacterial and host driven toxicity. J Neuroimmunol 184:45–52

Whitney CG, Pilishvili T, Farley MM, Schaffner W, Craig AS, Lynfield R, Nyquist AC, Gershman KA, Vazquez M, Bennett NM, Reingold A, Thomas A, Glode MP, Zell ER, Jorgensen JH, Beall B, Schuchat A (2006) Effectiveness of seven-valent pneumococcal conjugate vaccine against invasive pneumococcal disease: a matched case-control study. Lancet 368:1495–1502

Yoshimura A, Lien E, Ingalls RR, Tuomanen E, Dziarski R, Golenbock D (1999) Cutting edge: recognition of Gram-positive bacterial cell wall components by the innate immune system occurs via Toll-like receptor 2. J Immunol 163:1–5

Zughaier S, Agrawal S, Stephens DS, Pulendran B (2006) Hexa-acylation and KDO(2)-glycosylation determine the specific immunostimulatory activity of *Neisseria meningitidis* lipid A for human monocyte derived dendritic cells. Vaccine 24:1291–1297

Zughaier SM, Lindner B, Howe J, Garidel P, Koch MH, Brandenburg K, Stephens DS (2007) Physicochemical characterization and biological activity of lipooligosaccharides and lipid A from *Neisseria meningitidis*. J Endotoxin Res 13:343–357

Zughaier SM, Tzeng YL, Zimmer SM, Datta A, Carlson RW, Stephens DS (2004) *Neisseria meningitidis* lipooligosaccharide structure-dependent activation of the macrophage CD14/Toll-like receptor 4 pathway. Infect Immun 72:371–380

Zwijnenburg PJ, Van der Poll T, Florquin S, Akira S, Takeda K, Roord JJ, Van Furth AM (2003a) Interleukin-18 gene-deficient mice show enhanced defense and reduced inflammation during pneumococcal meningitis. J Neuroimmunol 138:31–37

Zwijnenburg PJ, Van der Poll T, Florquin S, Roord JJ, Van Furth AM (2003b) IL-1 receptor type 1 gene-deficient mice demonstrate an impaired host defense against pneumococcal meningitis. J Immunol 170:4724–4730

Toll-Like Receptors in Brain Abscess

Nilufer Esen and Tammy Kielian

Contents

Abstract Brain abscesses arise from a localized parenchymal infection, typically elicited by pyogenic bacteria such as *Staphylococcus aureus*. Despite improvements in detection and treatment strategies, brain abscesses continue to occur, with an increased prevalence in developing countries and immune-compromised patients. Adding to the seriousness of these infections is the recent emergence of antibiotic-resistant strains of bacteria, which are becoming more commonly associated with brain abscesses. Recent studies using a mouse experimental brain abscess model have revealed a complex role for Toll-like receptors (TLRs) in disease pathogenesis. Interestingly, TLR2 has limited impact on the innate immune response during the

N. Esen
Department of Neurology, University of Michigan Medical School, Ann Arbor, MI, USA

T. Kielian (✉)
Department of Pathology and Microbiology, University of Nebraska Medical Center, Omaha, NE, USA
e-mail: tkielian@unmc.edu

T. Kielian (ed.), *Toll-like Receptors: Roles in Infection and Neuropathology*, Current Topics in Microbiology and Immunology 336,
DOI 10.1007/978-3-642-00549-7_3,

acute stage of brain abscess formation induced by *S. aureus* but influences adaptive immunity. In contrast, mice deficient in MyD88, a central adapter molecule for the majority of TLRs in addition to the IL-1R and IL-18R, demonstrate severe defects in innate immunity coupled with exaggerated tissue destruction. It is envisioned that understanding the roles for TLRs in both resident CNS glia as well as infiltrating immune cells will provide insights into how the immune response to bacterial infection can be tailored to achieve effective pathogen destruction without inducing excessive bystander damage of surrounding noninfected brain parenchyma. A discussion of recent findings in this field is presented along with outstanding questions and the concept of a pathogen–necrosis–autoantigen triad for the amplification of TLR signaling is introduced.

Abbreviations

CNS	Central nervous system
CT	Computed tomography
GFAP	Glial fibrillary acidic protein
IFN-γ	Interferon-gamma
IL-1β	Interleukin-1 beta
IL-1R	Interleukin-1 receptor
IL-17	Interleukin 17
IL-17RA	IL-17 receptor A
IL-17RC	IL-17 receptor C
IL-18R	Interleukin-18 receptor
iNOS	Inducible nitric oxide synthase
KC	Keratinocyte chemoattractant
KO	Knockout
LOX-1	Lectin-like oxidized low-density lipoprotein receptor 1
LPS	Lipopolysaccharide
MAPK	Mitogen-activated protein kinase
MCP-1	Macrophage chemoattractant protein 1
MHC	Major histocompatibility complex
MIP-1α	Macrophage inflammatory protein-1 alpha
MIP-1β	Macrophage inflammatory protein-1 beta
MIP-2	Macrophage inflammatory protein 2
MRI	Magnetic resonance imaging
MSR	Macrophage scavenger receptor
MyD88	Myeloid differentiation primary-response protein 88
NF-κB	Nuclear factor kappa B
NOD2	Nucleotide-binding oligomerization domain containing 2
PAMP	Pathogen-associated molecular pattern
PGN	Peptidoglycan
PKC	Protein kinase C
PRR	Pattern recognition receptor

SPF	Specific pathogen-free
Th1	T helper 1
Th17	T helper 17
TLR	Toll-like receptor
TNF-α	Tumor necrosis factor alpha
WT	Wild type

1 Etiopathogenesis of Brain Abscess

Brain abscesses develop in response to localized infections in the parenchyma, leading to the formation of an encapsulated suppurative lesion. The main etiologic agents of brain abscess are the *Streptococcal* strains and *Staphylococcus aureus*, although a variety of organisms including Gram-negative bacteria and fungi have also been reported (Mathisen and Johnson 1997; Townsend and Scheld 1998; Bhand 2004; Prasad et al. 2006; Njambi et al. 2007; Kantarcioglu et al. 2008). Brain abscesses can form by bacterial perforation of the thin bony structures separating the brain from neighboring sites of chronic infection occurring in the paranasal sinuses, middle ear, or upper molars. Other routes include seeding of the brain with bacterial emboli originating from systemic sites of infection (i.e., endocarditis or septicemia) (Mathisen and Johnson 1997), where the frontal and temporal lobes are most commonly affected (McClelland et al. 1978; Carpenter et al. 2007), penetrating trauma to the head, or following neurosurgery (Tenney 1986; Schliamser et al. 1988). In addition, predisposing conditions such as congenital heart disease, hereditary hemorrhagic telangiectasia, or hemorrhagic or ischemic strokes have been associated with an increased incidence of brain abscess (Chen et al. 1995; Kumar 2000; Emmez et al. 2007; Sell et al. 2008). From a therapeutic standpoint, the diagnosis of brain abscess can often be delayed since patients typically present with nondescript clinical symptoms, including fever, poorly localized headache, and nausea. A differential diagnosis that includes brain abscess is often not made until patients begin to demonstrate neurological signs such as seizures, hemiparesis, and/or cognitive decline. By this time, a considerable amount of time has elapsed allowing the infection to disseminate and destroy a large portion of brain parenchyma. Therefore, if the diagnosis is significantly delayed and/or proper treatment is not administered, brain abscesses represent a significant medical problem, with mortality rates ranging from 30 to 80%. A large percentage of patients who recover from brain abscesses experience long-term complications, including seizures, loss of mental acuity, and focal neurological defects (Mathisen and Johnson 1997). The incidence of brain abscess is one in every 10,000 hospital admissions in the US of an infectious disease nature (Mathisen and Johnson 1997) and 1.3 per 100,000 persons worldwide (Nicolosi et al. 1991). For unknown reasons, the prevalence of brain abscess in males is 2–3 times more frequent compared to females (McClelland et al. 1978; Nicolosi et al. 1991; Kao et al. 2003). In underdeveloped countries, the disease incidence is higher and the source of infection typically originates from the ear or teeth due to poor sanitary conditions. In developed countries, improvements in health quality, advances in diagnostic tools, and treatment modalities improve infection outcome; however, the recent emergence of multi-drug-resistant microorganisms presents a new therapeutic challenge (Krcmery and Fedor-Freybergh 2007). Indeed, several case reports have described the isolation of methicillin-resistant *S. aureus* (MRSA) from brain abscesses, revealing the increased incidence of antibiotic-resistant bacteria in recent years (Smith et al. 1997; Khan et al. 2000; Roche et al. 2003; Jones et al. 2004). Moreover, immunosuppressive therapies employed for the treatment of tumors or following organ/bone marrow transplantation are associated with an increased prevalence of brain abscess formation often caused by commensal microorganisms (Calfee and Wispelwey 2000; Baddley et al. 2002; Hristea et al. 2007). Collectively, these issues highlight the need for further studies to understand host–pathogen interactions during brain abscess development.

2 Experimental Brain Abscess Model

Despite the continued incidence of brain abscesses, as demonstrated by the vast number of case reports in the literature, experimental models examining basic scientific aspects of this disease are limited. Although pioneering data were provided by experimental models of brain abscess in the rabbit (Falconer et al. 1943), dog (Britt et al. 1981) and rat (Winn et al. 1979; Flaris and Hickey 1992), studies describing the nature of the ensuing CNS and peripheral immune responses were rare or not detailed. Our laboratory has developed an experimental brain abscess model to elucidate the importance of host immune factors in disease pathogenesis (Kielian et al. 2001a,b; Baldwin and Kielian 2004; Kielian et al. 2004b), which was adapted from a previously published model in the rat (Flaris and Hickey 1992) and utilizes *S. aureus*, one of the main etiologic agents of brain abscess in humans (Roche et al. 2003; Jones et al. 2004) (Mathisen and Johnson 1997; Townsend and Scheld 1998). The mouse brain abscess model accurately reflects the course of disease progression in humans, providing an excellent model system to study immunological pathways influencing abscess pathogenesis and the effects of therapeutic agents on disease outcome.

At the histological level, brain abscess is typified by a sequential series of pathological changes that have been elucidated using the experimental rodent models described in detail below (Flaris and Hickey 1992; Kielian and Hickey 2000; Kielian et al. 2001a,b, 2004b). Staging of brain abscess in humans has been based on findings obtained during computed tomography (CT) or magnetic resonance imaging (MRI) scans. The early stage or early cerebritis occurs from days 1 to 3 and is typified by neutrophil accumulation, tissue necrosis, and edema. Microglial and astrocyte activation is also evident at this stage and persists throughout abscess development. The intermediate or late cerebritis stage occurs from days 4 to 9 and is associated with a predominant macrophage infiltrate. The final or capsule stage occurs from days 10 onward and is associated with lymphocytic infiltrates and the formation of a well-vascularized abscess wall, in effect sequestering the lesion and protecting the surrounding normal brain parenchyma from additional damage. It should be noted that these intervals represent general time frames that may deviate slightly depending on the colonizing pathogen and other host factors such as immune status or age. In addition to limiting the extent of infection, the immune response that is an essential part of abscess formation also destroys surrounding normal brain tissue. This is supported by findings in experimental models where lesion sites are greatly exaggerated compared to the localized nature of initial bacterial infection, reminiscent of an overactive immune response (Kielian et al. 2001a,b; Baldwin and Kielian 2004). This phenomenon is also observed in human brain abscess, where lesions can encompass a large portion of brain tissue, often spreading well beyond the initial focus of infection. Therefore, controlling the intensity and/or duration of anti-bacterial immune responses in the brain may allow for effective elimination of bacteria while minimizing damage to surrounding brain tissue. This may conceivably be achieved by regulating the extent of TLR activation during brain abscess development. For example, upon abscess sterilization, the presence of microbial debris such as cell wall fragments and/or bacterial DNA may serve as a continual trigger for TLR-mediated activation of resident glia and infiltrating immune cells. This would effectively culminate in prolonged cytokine/chemokine release that may exacerbate the destruction of normal surrounding brain parenchyma via bystander lysis. In effect, the brain immunologically "senses" an infection in the absence of viable bacteria, resulting in pathological inflammation. In addition, tissue necrosis that accompanies these infections likely liberates endogenous TLR ligands, in a so-called "pathogen-necrosis-autoantigen triad" that effectively amplifies TLR signaling to further exacerbate tissue damage. This concept is introduced in a later section of this chapter. However, these possibilities remain to be directly tested in the experimental brain abscess model.

3 Importance of TLRs in Glial Recognition of *S. aureus*

As previously mentioned, microglial and astrocyte activation ensues immediately following bacterial infection and persists throughout the various stages of brain abscess development. Both glial cell types are important for providing a first line of defense against microbes invading the CNS parenchyma. Astrocytes are strategically located at the blood–brain barrier, providing a mechanism to immediately sense pathogenic motifs. Based on its prevalence in human brain abscesses, we have focused our attention on characterizing the receptors responsible for eliciting glial activation and subsequent proinflammatory mediator production in response to *S. aureus*. We have primarily focused on the roles of TLR2 and the central adapter molecule MyD88 in these processes using primary glia isolated from TLR2 and MyD88 KO mice. A summary of the findings obtained with microglia and astrocytes is provided below.

3.1 TLR and Microglial Responses to S. aureus

Microglia are the resident professional mononuclear phagocyte population in the brain parenchyma, and function as the principal innate immune effector cells in the CNS (Aloisi 2001; Rivest 2003). Upon pathogen recognition, resting microglia transform into activated cells that migrate and accumulate at sites of injury (Gonzalez-Scarano and Baltuch 1999; Gonzalez-Scarano and Martin-Garcia 2005). Activated microglia express a range of genes related to inflammation, such as proinflammatory cytokines, proinflammatory enzymes, and adhesion molecules, which are dictated by the nature of the inflammatory stimulus (Gonzalez-Scarano and Baltuch 1999; Aloisi 2001; Block et al. 2007). The pattern of inflammatory molecule production can dictate leukocyte migration across the blood–brain barrier (Persidsky et al. 1999) and promote the effector functions of these newly recruited cells. Recent studies have demonstrated the presence of mRNA and/or protein expression of TLRs 1–9 in microglia (Dalpke et al. 2002; Rasley et al. 2002a; Kielian et al. 2002; Olson and Miller 2004), and have shown that TLR levels are modulated following exposure to bacterial pathogens (Rasley et al. 2002a; Kielian et al. 2002; Olson and Miller 2004; Esen and Kielian 2005). These findings are, perhaps, not surprising considering that microglia share the same myeloid lineage as macrophages and dendritic cells—important anti-bacterial effector and antigen-presenting cells.

One well-characterized PAMP of *S. aureus* is peptidoglycan (PGN), a major component of the bacterial cell wall (Dziarski 2003; Weber et al. 2003). Although recently scrutinized, PGN is still considered a potent TLR2 agonist (Dziarski and Gupta 2005). With regard to brain abscess, PGN is released during normal bacterial growth as well as from dying organisms within the necrotic environment that is typical of these infections. In addition, many antibiotics that are used to treat CNS Gram-positive infections enhance PGN release from the bacterial cell wall (van der Flier et al. 2003; Weber et al. 2003), liberating additional antigen to engage PRRs such as TLR2. Collectively, these findings indicate that PGN represents a PAMP of significant biological importance in brain abscess as well as other CNS Gram-positive infections. Therefore, our in vitro studies have centered on the responses of glial cells to PGN as well as intact *S. aureus*, the latter of which presents glia with a more complex antigenic milieu and multiple TLR ligands (i.e., lipoproteins, bacterial DNA).

Our results have demonstrated that microglia respond to both intact *S. aureus* and its cell wall component PGN robustly. Specifically, exposure to both stimuli led to a dose- and time-dependent induction of the proinflammatory cytokines IL-1β, TNF-α, IL-12 p40, and several chemokines including CXCL1 (KC), CXCL2 (MIP-2), CCL2 (MCP-1), CCL3 (MIP-1α), and CCL4 (MIP-1β) (Kielian et al. 2002, 2004a, 2005a). The importance of microglia in the early host response to infection in brain abscess is suggested by the fact that proinflammatory mediator production is detected within 1–3 h following the initial *S. aureus* infection, well before the significant accumulation of peripheral immune cell infiltrates (Kielian and Hickey 2000; Esen et al. 2004). Moreover, both heat-inactivated *S. aureus* and PGN are capable of inducing microglial MHC Class II, CD40, CD80, and CD86 expression (Kielian et al. 2002, 2004a; Esen and Kielian 2007), similar to what has been described for microglia in response to the Gram-negative bacterial cell wall product lipopolysaccharide (LPS) and interferon-γ (IFN-γ) (Frei et al. 1994; Xu and Ling 1995; Menendez Iglesias et al. 1997; Aloisi et al. 1998; O'Keefe et al. 2001). The ability of *S. aureus* to augment the expression of receptors that are important for antigen presentation suggests that the ability of microglia to present bacterial peptides to antigen-specific T cells may be greatly enhanced following *S. aureus* exposure. The effects of *S. aureus* and PGN on microglial CD40, CD80, CD86, and MHC Class II expression may be either a direct consequence of bacterial stimulation or an indirect one via the autocrine/paracrine action of cytokines produced by activated microglia.

An analysis of the importance of TLR2 in mediating microglial responses to *S. aureus* revealed some interesting findings. Namely, primary microglia isolated from TLR2 KO mice were nonresponsive to *S. aureus*-derived PGN, demonstrating a pivotal role for TLR2-dependent signals in eliciting proinflammatory mediator release from microglia in response to this PAMP (Kielian et al. 2005a). In contrast, microglial activation following treatment with intact *S. aureus* was only marginally affected by TLR2 loss, with the majority of mediators expressed at roughly equivalent levels between TLR2 KO and WT primary microglia. Interestingly, we consistently observed elevated IL-12 p40 production by TLR2 KO microglia in response to intact *S. aureus*, suggesting dysregulated signaling pathways (Kielian et al. 2005a). Heightened IL-12 p40 expression was never detected in TLR2 KO microglia following PGN treatment, indicating antigenic specificity in cytokine regulation. The mechanism(s) responsible for this phenomenon are not known but are under investigation in our laboratory.

Unlike what was observed with TLR2, MyD88 expression was essential for eliciting proinflammatory mediator production from microglia in response to both intact *S. aureus* as well as PGN. This finding strongly suggests the involvement of alternative TLRs that recognize additional PAMPs present in *S. aureus* and/or amplification of signaling networks through the autocrine/paracrine actions of cytokines that utilize MyD88, such as the IL-1R and IL-18R. The importance of MyD88 in CNS resident cells has recently been demonstrated using radiation bone marrow chimera mice (Garg et al., 2009). Specifically, cytokine expression was restored to wild-type levels during brain abscess development only when MyD88 was present in the CNS. The absence of MyD88 in the CNS compartment, with expression restricted to infiltrating immune cells, was not sufficient to restore cytokine levels to that observed in wild-type mice. Although we cannot ascertain whether MyD88 is pivotal in microglia, astrocytes, and/or other CNS resident cell types, these data do indicate that CNS MyD88 is essential for generating a potent inflammatory response during brain abscess development.

3.2 Astrocytes and TLR2-dependent Recognition of *S. aureus*

Besides microglia, astrocytes can also participate in the immune response against CNS pathogens, in addition to their traditional roles in influencing CNS development and homeostasis. Indeed, activated astrocytes have been recognized as being a major source of inflammatory chemokines in response to diverse stimuli (Dong and Benveniste 2001; Farina et al. 2007). Several reports have provided direct evidence for the presence of TLRs in astrocytes (Bsibsi et al. 2002; Bowman et al. 2003; Esen et al. 2004; Carpentier et al. 2005; Jack et al. 2005); however, the number of TLRs expressed is much lower compared to microglia, which may be explained by the fact that microglia are more classical innate immune effector cells. Primary murine astrocytes constitutively express low levels of TLR2, TLR3, TLR5 and TLR9, with the expression of each TLR homolog rapidly upregulated following exposure to its cognate bacterial ligand (Bowman et al. 2003; Carpentier et al. 2005). Although some groups have reported that astrocytes express TLR4 (references 1–3 below), we and others have been unable to demonstrate receptor expression (references 4–6 below and Kielian, unpublished observations). These discrepancies may be due to signals arising from residual microglial contamination since these cells are notorious for hiding beneath astrocytic monolayers and remains an issue to be resolved.

With regard to *S. aureus*, astrocytes appear to rely heavily on TLR2 for signaling cell activation and subsequent proinflammatory mediator release. This was demonstrated by the finding that TLR2 KO astrocytes were relatively nonresponsive to either PGN or intact *S. aureus* (Esen et al. 2004), the latter observation being a major discriminating factor between astrocytes and microglia that utilize alternative PRRs for maximal recognition of intact bacteria. Specifically, a wide array of proinflammatory cytokines and chemokines were attenuated in TLR2 and MyD88 KO astrocytes despite the fact that bacterial internalization was not affected by either molecule (Esen et al. 2004) (Esen and Kielian, manuscript in preparation). As mentioned earlier, MyD88 plays an important role in the CNS compartment in eliciting a protective immune response during the acute stage of brain abscess development (Garg et al., 2009). It is quite probable that astrocytes, in concert with microglia, contribute to this process. Since astrocytes represent a more robust source of chemokines compared to microglia, the loss of MyD88 in the former may have drastic implications on the recruitment of immune cells from the periphery into the infected CNS.

3.3 Cytokine Regulation of TLR2 Expression in Glia

We previously reported that exposure of microglia to *S. aureus* led to elevated TLR2 expression (Kielian et al. 2005a). However, it remained uncertain as to whether this phenomenon resulted from a direct effect of pathogen exposure or indirectly through the autocrine/paracrine action of cytokines released from activated cells. Indeed, microglia produce several inflammatory mediators following *S. aureus* stimulation, including TNF-α and IL-1β (Kielian et al. 2002, 2004a, 2005a). We recently demonstrated that the proinflammatory cytokine TNF-α enhances TLR2 expression in microglia, whereas interleukin-1β (IL-1β) had no significant effect (Syed et al. 2007). Strong evidence implicating TNF-α as a pivotal mediator for augmenting TLR2 expression was demonstrated by the inability of TNF-α KO microglia to upregulate TLR2 following *S. aureus* stimulation (Syed et al. 2007). Direct in vivo evidence to support this concept was provided by the finding that TLR2 expression was significantly attenuated during brain abscess development in TNF-α KO mice. In addition, the TNF-α-dependent induction of TLR2 expression in microglia was mediated by a NF-κB-dependent pathway, whereas PKC or p38 MAPK signals were not involved (Syed et al. 2007). Similar findings were obtained in astrocytes, where TNF-α was shown to act in an autocrine/paracrine fashion to augment TLR2 expression (Phulwani et al. 2008). The ability of TNF-α to increase TLR2 levels was mediated by NF-κB signaling since disparate inhibitors of this pathway were capable of interfering with TLR2 induction (Phulwani et al. 2008). Collectively, these results suggest that TNF-α likely plays an important role in amplifying the immune responses of both microglia and astrocytes upon *S. aureus* exposure, in part by augmenting TLR2 expression.

3.4 *Potential Contributions of Other TLRs to S. aureus Recognition by Glia*

Our findings with TLR2 KO primary microglia suggested that other receptors are also involved in maximal *S. aureus* recognition. Indeed, this functional redundancy is not surprising since bacterial pathogens have the potential for devastating consequences in a tissue that has limited regenerative capacity such as the CNS. Possible alternative receptors could be TLR9, which recognizes nonmethylated CpG DNA motifs of microbes, or other cytoplasmic PRRs which may encounter PAMPs such as the cytosolic NOD2 protein which binds PGN (Inohara et al. 2003; Girardin et al. 2003a,b). Alternatively, phagocytic scavenger receptors such as macrophage scavenger receptor type AI/AII (MSR) and lectin-like oxidized low-density lipoprotein receptor (LOX)-1 may play a role. Since the expression of LOX-1 is significantly increased in microglia following *S. aureus* exposure in vitro and in brain abscesses in vivo (Kielian et al. 2005a,b), we suggest that scavenger receptors may be utilized by microglia to phagocytize and inactivate bacteria. More supporting evidence came with data showing that bacterial phagocytosis is opsonin-independent and does not require TLRs, since microglia lacking the central adapter molecule MyD88 were still capable of phagocytizing *S. aureus* (Esen and Kielian 2006, and unpublished results).

The implications of glial activation in the context of brain abscess are likely several-fold. First, parenchymal microglia and astrocytes are likely involved in the initial recruitment of professional bactericidal phagocytes into the CNS through their elaboration of chemokines and proinflammatory cytokines. Second, microglia exhibit modest *S. aureus* bactericidal activity in vitro, suggesting that they may also participate in the initial containment of bacterial replication in the CNS. However, their bactericidal activity in vitro is not comparable to that of neutrophils or macrophages, suggesting that this activity may not be a major effector mechanism for microglia during acute infection. Third, activated microglia have the potential to influence the type and extent of anti-bacterial adaptive immune responses through their upregulation of MHC Class II and co-stimulatory molecule expression. Finally, if glial activation persists in the context of ongoing inflammation, the continued release of proinflammatory mediators could damage surrounding normal brain parenchyma. Indeed, inappropriate glial activation has been implicated in several CNS diseases, including multiple sclerosis and its animal model experimental autoimmune encephalomyelitis, as well as Alzheimer's disease (Benveniste 1997; McGeer and McGeer 2002).

4 Impact of TLR Signaling on Immune Responses During Brain Abscess Development

TLR2 has been shown to play an important role in the host immune response to Gram-positive bacterial infections in the periphery (Takeuchi et al. 2002) and, to some extent, this receptor dictates the ensuing host anti-bacterial response in *Streptococcus pneumoniae* meningitis (Echchannaoui et al. 2002; Koedel et al. 2003) (see the chapter "Toll-Like Receptors in Bacterial Meningitis" of this volume). However, prior to our studies, the functional role of TLR2 in the context of a CNS parenchymal infection, such as brain abscess, had not been examined and was thought to differ from that of meningitis based upon the highly focal nature of lesions in the former.

4.1 Influence of TLRs on Innate Immunity in Brain Abscesses

We first evaluated the expression of numerous proinflammatory mediators previously determined to be pivotal for the host immune response during the acute phase of brain abscess development to ascertain whether defects in CNS bacterial recognition were evident in TLR2-deficient animals. The kinetics of proinflammatory mediator production, including TNF-α, IL-1β, CXCL2 and inducible nitric oxide synthase (iNOS), were delayed in TLR2 KO mice compared to WT animals, with lower levels of mediators detected in the former during the acute stage of disease (Kielian et al. 2005a,b). However, the reduction in these inflammatory mediators was transient, with levels in TLR2 KO mice recovering to those detected in WT animals over time, and as such, no striking phenotype was observed in TLR2 KO mice, much to our surprise. However, one dramatic difference was consistently observed in brain abscesses of TLR2 KO animals, namely elevated IL-17 expression (Kielian et al. 2005b). The potential biological impact of increased IL-17 levels in TLR2 KO mice will be discussed in a later section describing the interface of TLRs with the adaptive immune response. In addition, TLR2 did not play a significant role in controlling the extent of infection in brain abscess, with similar bacterial titers observed between TLR2 KO and WT animals, suggesting receptor redundancy for *S. aureus* neutralization in the CNS (Kielian et al. 2005a,b). Interestingly, the inflammatory phenotype detected in TLR2 KO mice was nearly identical to that observed in CD14 KO animals (Kielian, unpublished observations), strongly suggesting that these two receptors may cooperate in a multireceptor complex to facilitate pathogen recognition and the subsequent shaping of the inflammatory milieu during the acute stage of infection. These results were in accordance with our in vitro studies, implying a role for alternative receptor(s) in bacterial recognition and subsequent activation of glial cells (Esen et al. 2004; Kielian et al. 2005a). It is worth noting that our findings with TLR2 KO mice in the experimental brain abscess model (Kielian et al. 2005b) are at odds with a recent report by Stenzel et al. (2008). Specifically, Stenzel et al. noted that brain abscess size was exacerbated in TLR2 KO mice; however, bacterial burdens were only marginally increased (Stenzel et al. 2008). Similar to our earlier report (Kielian et al. 2005b), this group also observed that the induction of proinflammatory mediator expression was delayed in TLR2 KO animals but eventually achieved levels equivalent to WT mice (Stenzel et al. 2008).

In contrast to studies with TLR2 KO mice, MyD88 plays an essential role during the early phase of brain abscess. In particular, MyD88 loss led to a significant increase in mortality rates, which correlated with the complete disruption of normal brain tissue architecture in the infected parenchyma (Kielian et al. 2007). Concomitant with these outcomes was the inability to recruit significant numbers of neutrophils and macrophages into the brain parenchyma of MyD88 KO animals, the former representing a leukocyte population that we had previously demonstrated to be essential for *S. aureus* containment during the acute stage of brain abscess (Kielian et al. 2001a). Finally, MyD88 KO mice displayed significant decreases in the expression of numerous proinflammatory cytokines and chemokines, which likely contributed, in part, to the impaired leukocyte recruitment observed in KO animals (Kielian et al. 2007). The dramatic phenotype observed during the early stage of brain abscess formation in MyD88 KO mice is likely influenced by the loss of multiple TLRs as well as signaling via the IL-1R and/or IL-18R. Indeed, earlier studies by our group demonstrated that IL-1 KO mice displayed more severe infection, as typified by increased bacterial burdens and alterations in inflammatory mediator production during brain abscess development (Kielian et al. 2004b). The relative role of IL-18 signaling is uncertain since no one has yet investigated the importance of IL-18 in the experimental brain abscess model.

MyD88-dependent signals also played an important role in regulating the extent of parenchymal damage during the acute stage of brain abscess development. Indeed, brain abscesses were grossly disseminated in MyD88 KO mice and often encompassed the entire injected hemisphere, whereas lesions were contained and well demarcated in WT animals at all time points examined (Kielian et al. 2007).

A recent study by Stenzel et al. has demonstrated that the astrocytic intermediate filament protein GFAP is important for providing a barrier to restrict brain abscess expansion throughout the parenchyma (Stenzel et al. 2004). Of note is the finding that in our studies, GFAP immunoreactivity was enhanced in MyD88 KO mice (Kielian, unpublished observations), yet these animals had diffuse lesions with ill-defined borders. This result may reflect an attempt by astrocytes to sequester the lesion for effective bacterial containment; however, in the face of the dramatic defects in CNS innate immunity in MyD88 KO mice, this attempt is futile. The dramatic pathology observed in brain abscesses of MyD88 KO mice presents an excellent model system to test the efficacy of various novel antimicrobial therapies since it is reminiscent of a worst-case scenario clinically.

Neutrophil and macrophage accumulation into brain abscesses was significantly attenuated in MyD88 KO mice compared to WT animals (Kielian et al. 2007). Importantly, this correlated with a dramatic decrease in the production of numerous chemokines in brain abscesses of the former. The number of abscess-associated neutrophils in MyD88 KO mice was more significantly affected compared to macrophages, which agreed with the finding that the expression of neutrophil chemokines (i.e., CXCL1, CXCL2) was more depressed compared to macrophage chemoattractants (i.e., CCL2, CCL3). Not only were the numbers of abscess-associated neutrophils and macrophages reduced in brain abscesses of MyD88 KO mice, but their activation status was also impaired. This was reflected by lower levels of several cytokines and chemokines including IL-6, TNF-α, CCL3, and CXCL10 produced by these cell types recovered from lesions of MyD88 KO mice. Collectively, these findings highlight the importance of MyD88-dependent signals in inducing the host innate immune response during *S. aureus* infection in the brain.

One intriguing and unexpected finding that surfaced during the course of our studies with MyD88 KO mice was the fact that bacterial burdens were equivalent in MyD88 KO and WT animals despite the dramatic inhibition of CNS innate immune responses in the former. This finding suggests that alternative mechanisms are responsible for controlling bacterial replication in the CNS parenchyma, since no role for TLRs was evident in our studies. Since TLRs are not phagocytic receptors, it may be expected that alternative PRRs are responsible for pathogen uptake and neutralization (Henneke et al. 2002; Underhill and Gantner 2004). Some candidate receptors that may participate in regulating *S. aureus* burdens in the CNS parenchyma include members of the scavenger receptor family as well as opsonic receptors, including Fc and complement receptors (Husemann et al. 2002; Peiser et al. 2002). Another possibility is that the extensive tissue damage observed in MyD88 KO mice leads to cell death via a mechanism that is independent of bacterial burdens per se. Perhaps with a longer infection period differences in bacterial burdens between MyD88 KO and WT mice would have become evident; however, this analysis was not feasible in our studies since the majority of KO animals did not survive beyond 48 h post-infection. An alternative explanation may be that the defective host immune response in MyD88 KO mice led to altered selective pressures on *S. aureus* that could conceivably modify the organism's virulence factor expression profile. In this case, bacteria may have switched on a cohort of genes that favored invasion and dissemination. We are currently investigating this possibility by performing transcriptional profiling of genes expressed by *S. aureus* in MyD88 KO and WT mice using Affymetrix GeneChips®.

4.2 Relationship Between TLR2 and Adaptive Immunity in Brain Abscesses

Innate and adaptive immunity are linked, and recent evidence indicates that TLR-dependent signaling leads to the initiation of adaptive immune responses (Hoebe et al. 2004; Pasare and Medzhitov 2005). Indeed, our earlier studies demonstrating lymphocyte influx and generation of *S. aureus*-reactive lymphocytes in the brain abscess model (Kielian and Hickey 2000; Baldwin and Kielian 2004) as well as the induction of molecules involved in antigen presentation (Kielian et al. 2002, 2004a; Esen and Kielian 2007) have provided clues to the connection between innate and adaptive immunity in brain abscesses. One apparent link between the two processes was discovered when we detected elevated IL-17 levels in TLR2 KO mice during the early stages of primary infection (Kielian et al. 2005b). In this scenario, attributing cytokine production to memory T cells appeared unusual since animals had not been infected earlier with *S. aureus*. However, we have previously reported the presence of a population of *S. aureus*-reactive T cells in the spleens of "naïve" mice under non-SPF conditions (Baldwin and Kielian 2004). The origin of these *S. aureus*-specific lymphocytes is not known but may have arisen from routine pathogen exposure, since *S. aureus* is ubiquitous in nature, or from the restimulation of cross-reactive lymphocytes that are specific for a highly conserved epitope expressed on another bacterial pathogen. Therefore, it is conceivable that "endogenous" *S. aureus*-reactive memory T cells are recruited into the CNS during the early stages of brain abscess development and are responsible for subsequent IL-17 production. Indeed, we have recently demonstrated that the numbers of both $CD4^+$ and $CD8^+$ T cell infiltrates are enhanced in brain abscesses of TLR2 KO mice compared to WT animals (Nichols et al., 2009). These findings are in agreement with a recent report by Stenzel et al. in the experimental brain abscess model (Stenzel et al. 2008), providing further evidence that T cell dynamics are indeed altered in the infected CNS of mice lacking TLR2. It is worth noting that elevated Th17 influx in TLR2 KO animals appears to represent a nontraditional pathway since, although Th17 cells are present in the CNS of infected WT mice, they do not dominate the adaptive response as Th1 cells producing IFN-γ are also a significant component of the T cell infiltrate (Nichols et al., 2009).

We have recently found that primary microglia and astrocytes express for IL-17R isoforms, namely IL-17RA and IL-17RC, although not at equivalent levels (Nichols et al., 2009). Specifically, IL-17RA and IL-17RC were more highly expressed in microglia compared to astrocytes, whereas neither receptor was further modulated by bacterial activation. Evidence to support the idea that increased IL-17 production in TLR2 KO animals represents a compensatory mechanism to counteract the observed delay in the production of neutrophil-attracting chemokines came with the observation that IL-17 induced the expression of the neutrophil chemoattractant CXCL2 in astrocytes (Nichols et al., 2009), in accordance with several studies that report IL-17 as a potent stimulus for the production of neutrophil chemokines (Witowski et al. 2000; Maertzdorf et al. 2002; Ruddy et al. 2004a). In addition, IL-17 has been shown to be pivotal in the establishment of antimicrobial immunity, since IL-17 KO mice are more susceptible to systemic bacterial infections (Ye et al. 2001; Chung et al. 2002, 2003).

Since IL-17 has been reported to synergize with other proinflammatory cytokines such as TNF-α and IL-1β in various cell types (Chabaud et al. 2001; Maertzdorf et al. 2002; Ruddy et al. 2004b; Dong 2008), we examined whether this relationship would hold true for glia. Although TNF-α and IL-1β alone were capable of eliciting the production of a few inflammatory mediators by glia, the addition of IL-17 did not significantly alter the release of these factors (Nichols et al., 2009). Collectively, the limited response of glial cells to IL-17 stimulation suggests that the primary target of increased IL-17 levels in brain abscesses of TLR2 KO mice may be stimulation of alternative parenchymal cells and/or peripheral immune cells infiltrating the CNS. Evidence to support the latter has been provided by recent studies from our laboratory demonstrating synergistic effects of IL-17 in combination with TNF-α in primary macrophages. These cells are also relevant to brain abscess pathogenesis since macrophages represent a significant constituent of the inflammatory infiltrate in response to infection (Kielian 2004). Importantly, only a select number of mediators were potentiated by IL-17 in the presence of TNF-α (i.e., CXCL2, CCL2), whereas others were not affected (i.e., IL-1β, IL-12). Interestingly, we failed to detect any synergy with IL-17 and IL-1β regardless of the concentrations examined, revealing specificity in the responses obtained. We cannot discount additional roles for IL-17 in other effector responses of microglia and astrocytes including phagocytosis, chemotaxis and/or increased responsiveness to other inflammatory signals. These remain areas of continued investigation in our laboratory.

4.3 Concept of a "Pathogen–Necrosis–Autoantigen Triad"

During brain abscess development, the resultant immune response (mediated, in part, via TLRs) leads not only to pathogen destruction but also bystander damage to surrounding CNS parenchyma by necrotic cell death. Further complicating the issue is the fact that bacteria express numerous virulence factors that can directly lead to cell death/lysis of mammalian cells. One example is the hemolysins produced by *S. aureus* that form transmembrane pores in mammalian cells leading to cell destruction by osmotic lysis. The end result of this necrotic damage is the release of self antigens, many of which would not typically be exposed to the extracellular milieu and available for uptake by antigen-presenting cells. Therefore, it is also possible that during brain abscess evolution TLRs play dual roles in ligand recognition. In addition, one is to facilitate the initial response to the inciting pathogen. Upon tissue destruction, TLRs may also recognize newly liberated self antigens as a result of necrosis, a series of events we have coined a "pathogen–necrosis–autoantigen triad" (Fig. 1). In this case, newly liberated self antigens may serve as direct triggers for TLRs and/or be internalized by antigen-presenting cells in the vicinity of the abscess (i.e., macrophages, dendritic cells, and microglia), whereupon they could conceivably stimulate autoimmune reactions. This "pathogen–necrosis–autoantigen triad" shares similarities with the proposed "fertile field"

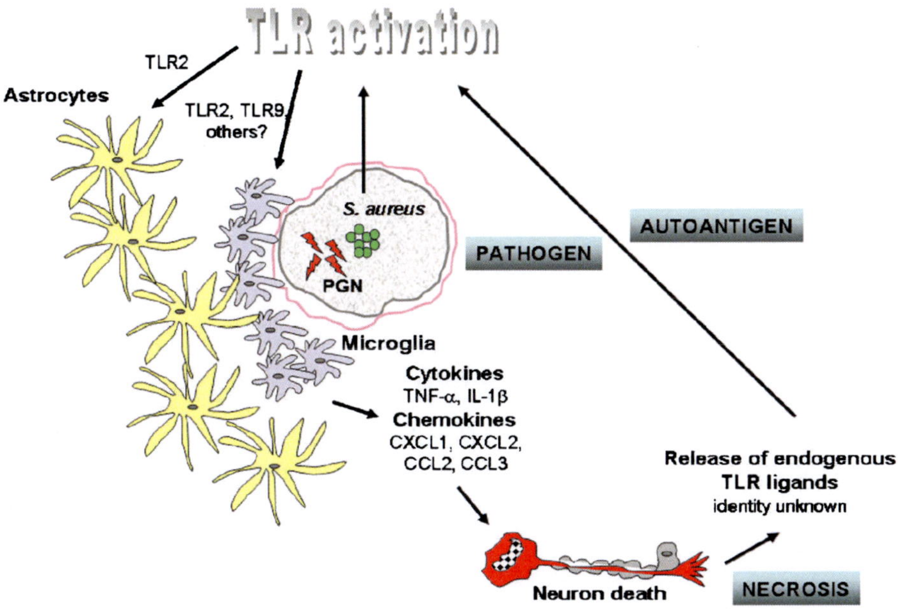

Fig. 1 Proposed pathogen-necrosis-autoantigen triad. With regard to brain abscesses induced by *S. aureus*, the pathogen induces microglia and astrocyte activation via engagement of several Toll-like receptors (TLRs) in addition to other receptors to elicit pro-inflammatory mediator release to combat the infection. However, this response also leads to extensive necrosis of the infected parenchyma, which is also facilitated by virulence factors produced by the bacterium. Necrosis leads to the release of self antigens (autoantigens) that would not typically be encountered in the extracellular milieu. It is plausible that some of these autoantigens (as of yet to be defined) serve as further triggers for TLR activation, effectively exacerbating inflammation and further bystander destruction to surrounding uninfected brain parenchyma

concept for autoimmunity, involving molecular mimicry in concert with bystander activation (von Herrath et al. 2003). However, it is imperative to point out that there is no evidence to date, either experimental or clinical, to indicate that patients who have recovered from brain abscess exhibit increased prevalence of subsequent autoimmune disease, although it is likely that no one has investigated this relationship with retrospective studies. Although it is clear that self antigens are liberated during brain abscess development as a result of necrosis, what is not certain is whether this primes an individual for downstream autoimmunity. It could very well be that the liberation of self antigen during disease serves as a trigger for further TLR activation (via endogenous TLR agonists), which merely contributes to the inflammatory foci without inducing long-lasting overt consequences.

5 Summary

Despite recent studies investigating the importance of TLR2 and MyD88 in regulating host immune responses during brain abscess development, several outstanding questions remain. For example, studies with TLR2 KO microglia and in the experimental brain abscess model have clearly demonstrated the presence of alternative redundant pathways for bacterial recognition. This was initially unexpected given the importance of TLR2 in *S. aureus* sepsis (Takeuchi et al. 2000); however, upon closer examination this may be a dose-dependent effect, since severe pathology was not observed systemically unless TLR2 KO mice were challenged with high numbers of bacteria (Takeuchi et al. 2000). Nonetheless, the identities of alternative receptors involved in *S. aureus* recognition in the CNS remain to be defined. Another unresolved issue is the functional importance of IL-17 during brain abscess development and, most importantly, whether IL-17 neutralization will reverse the proposed compensatory effect of elevated cytokine expression in brain abscesses of TLR2 KO mice. These are questions that are currently under investigation in our laboratory. Collectively, the TLR family of molecules likely influences a wide array of inflammatory responses during the course of brain abscess development, impacting both the innate and adaptive arms of the immune system.

Acknowledgments This work was supported by the NIH National Institute of Neurological Disorders and Stroke (RO1s NS055385, NS40730, and NS053487 to T.K).

References

Aloisi F (2001) Immune function of microglia. Glia 36:165–179

Aloisi F, Ria F, Penna G, Adorini L (1998) Microglia are more efficient than astrocytes in antigen processing and in Th1 but not Th2 cell activation. J Immunol 160(10):4671–4680

Baddley JW, Salzman D, Pappas PG (2002) Fungal brain abscess in transplant recipients: epidemiologic, microbiologic, and clinical features. Clin Transplant 16:419–424

Baldwin AC, Kielian T (2004) Persistent immune activation associated with a mouse model of *Staphylococcus aureus*-induced experimental brain abscess. J Neuroimmunol 151:24–32

Benveniste EN (1997) Role of macrophages/microglia in multiple sclerosis and experimental allergic encephalomyelitis. J Mol Med 75:165–173

Bhand AA (2004) Brain abscess—diagnosis and management. J Coll Physicians Surg Pak 14:407–410

Block ML, Zecca L, Hong JS (2007) Microglia-mediated neurotoxicity: uncovering the molecular mechanisms. Nat Rev Neurosci 8:57–69

Bowman CC, Rasley A, Tranguch SL, Marriott I (2003) Cultured astrocytes express toll-like receptors for bacterial products. Glia 43:281–291

Britt RH, Enzmann DR, Yeager AS (1981) Neuropathological and computerized tomographic findings in experimental brain abscess. J Neurosurg 55:590–603

Bsibsi M, Ravid R, Gveric D, van Noort JM (2002) Broad expression of Toll-like receptors in the human central nervous system. J Neuropathol Exp Neurol 61(11):1013–1021

Calfee DP, Wispelwey B (2000) Brain abscess. Semin Neurol 20:353–360

Carpenter J, Stapleton S, Holliman R (2007) Retrospective analysis of 49 cases of brain abscess and review of the literature. Eur J Clin Microbiol Infect Dis 26:1–11

Carpentier PA, Begolka WS, Olson JK, Elhofy A, Karpus WJ, Miller SD (2005) Differential activation of astrocytes by innate and adaptive immune stimuli. Glia 49:360–374

Chabaud M, Page G, Miossec P (2001) Enhancing effect of IL-1, IL-17, and TNF-alpha on macrophage inflammatory protein-3alpha production in rheumatoid arthritis: regulation by soluble receptors and Th2 cytokines. J Immunol 167:6015–6020

Chen ST, Tang LM, Ro LS (1995) Brain abscess as a complication of stroke. Stroke 26: 696–698

Chung DR, Chitnis T, Panzo RJ, Kasper DL, Sayegh MH, Tzianabos AO (2002) CD4 + T cells regulate surgical and postinfectious adhesion formation. J Exp Med 195:1471–1478

Chung DR, Kasper DL, Panzo RJ, Chitnis T, Grusby MJ, Sayegh MH, Tzianabos AO (2003) CD4 + T cells mediate abscess formation in intra-abdominal sepsis by an IL-17-dependent mechanism. J Immunol 170:1958–1963

Dalpke AH, Schafer MK, Frey M, Zimmermann S, Tebbe J, Weihe E, Heeg K (2002) Immunostimulatory CpG-DNA activates murine microglia. J Immunol 168:4854–4863

Dong C (2008) TH17 cells in development: an updated view of their molecular identity and genetic programming. Nat Rev Immunol 8:337–348

Dong Y, Benveniste EN (2001) Immune function of astrocytes. Glia 36:180–190

Dziarski R (2003) Recognition of bacterial peptidoglycan by the innate immune system. Cell Mol Life Sci 60:1793–1804

Dziarski R, Gupta D (2005) *Staphylococcus aureus* peptidoglycan is a toll-like receptor 2 activator: a reevaluation. Infect Immun 73:5212–5216

Echchannaoui H, Frei K, Schnell C, Leib SL, Zimmerli W, Landmann R (2002) Toll-like receptor 2-deficient mice are highly susceptible to *Streptococcus pneumoniae* meningitis because of reduced bacterial clearing and enhanced inflammation. J Infect Dis 186:798–806

Emmez H, Borcek AO, Dogulu F, Ceviker N (2007) Ischemic stroke complicated by a brain abscess: a case report and review of the literature. Turk Neurosurg 17:48–54

Esen N, Kielian T (2005) Recognition of *Staphylococcus aureus*-derived peptidoglycan (PGN) but not intact bacteria is mediated by CD14 in microglia. J Neuroimmunol 170:93–104

Esen N, Kielian T (2006) Central role for MyD88 in the responses of microglia to pathogen-associated molecular patterns. J Immunol 176:6802–6811

Esen N, Kielian T (2007) Effects of low dose GM-CSF on microglial inflammatory profiles to diverse pathogen-associated molecular patterns (PAMPs). J Neuroinflammation 4:10

Esen N, Tanga FY, DeLeo JA, Kielian T (2004) Toll-like receptor 2 (TLR2) mediates astrocyte activation in response to the Gram-positive bacterium *Staphylococcus aureus*. J Neurochem 88:746–758

Falconer MA, McFarlan, AM, Russell, DS (1943) Experimental brain abscesses in the rabbit. B J Surg 30:245–260

Farina C, Aloisi F, Meinl E (2007) Astrocytes are active players in cerebral innate immunity. Trends Immunol 28:138–145

Flaris NA, Hickey WF (1992) Development and characterization of an experimental model of brain abscess in the rat. Am J Pathol 141:1299–1307

Frei K, Lins H, Schwerdel C, Fontana A (1994) Antigen presentation in the central nervous system. The inhibitory effect of IL-10 on MHC class II expression and production of cytokines depends on the inducing signals and the type of cell analyzed. J Immunol 152:2720–2728

Garg S, Nichols JR, Esen N, Liu S, Phulwani NK, Syed MMd, Wood WH, Zhang Y, Becker KG, Aldrich A, Kielian T (2009) MyD88 expression in the CNS is pivotal for eliciting protective immunity in brain abscesses. ASN NEURO doi:10.1042/AN20090004

Girardin SE, Travassos LH, Herve M, Blanot D, Boneca IG, Philpott DJ, Sansonetti PJ, Mengin-Lecreulx D (2003a) Peptidoglycan molecular requirements allowing detection by Nod1 and Nod2. J Biol Chem 278:41702–41708

Girardin SE, Boneca IG, Viala J, Chamaillard M, Labigne A, Thomas G, Philpott DJ, Sansonetti PJ (2003b) Nod2 is a general sensor of peptidoglycan through muramyl dipeptide (MDP) detection. J Biol Chem 278:8869–8872

Gonzalez-Scarano F, Baltuch G (1999) Microglia as mediators of inflammatory and degenerative diseases. Annu Rev Neurosci 22:219–240

Gonzalez-Scarano F, Martin-Garcia J (2005) The neuropathogenesis of AIDS. Nat Rev Immunol 5:69–81

Henneke P, Takeuchi O, Malley R, Lien E, Ingalls RR, Freeman MW, Mayadas T, Nizet V, Akira S, Kasper DL, Golenbock DT (2002) Cellular activation, phagocytosis, and bactericidal activity against group B streptococcus involve parallel myeloid differentiation factor 88-dependent and independent signaling pathways. J Immunol 169:3970–3977

Hoebe K, Janssen E, Beutler B (2004) The interface between innate and adaptive immunity. Nat Immunol 5:971–974

Hristea I, Bunnapradist S, Peng A, Puliyanda D, Vo A, Jordan SC (2007) The onset of rapidly progressive neurologic deterioration after a brief gastrointestinal illness in a renal allograft recipient. Transpl Infect Dis 9:142–147

Husemann J, Loike JD, Anankov R, Febbraio M, Silverstein SC (2002) Scavenger receptors in neurobiology and neuropathology: their role on microglia and other cells of the nervous system. Glia 40:195–205

Inohara N, Ogura Y, Fontalba A, Gutierrez O, Pons F, Crespo J, Fukase K, Inamura S, Kusumoto S, Hashimoto M, Foster SJ, Moran AP, Fernandez-Luna JL, Nunez G (2003) Host recognition of bacterial muramyl dipeptide mediated through NOD2. Implications for Crohn's disease. J Biol Chem 278:5509–5512

Jack CS, Arbour N, Manusow J, Montgrain V, Blain M, McCrea E, Shapiro A, Antel JP (2005) TLR signaling tailors innate immune responses in human microglia and astrocytes. J Immunol 175:4320–4330

Jones ME, Draghi DC, Karlowsky JA, Sahm DF, Bradley JS (2004) Prevalence of antimicrobial resistance in bacteria isolated from central nervous system specimens as reported by U.S. hospital laboratories from 2000 to 2002. Ann Clin Microbiol Antimicrob 3:3

Kantarcioglu AS, Guarro J, de Hoog GS (2008) Central nervous system infections by members of the *Pseudallescheria boydii* species complex in healthy and immunocompromised hosts: epidemiology, clinical characteristics and outcome. Mycoses 51:275–290

Kao PT, Tseng HK, Liu CP, Su SC, Lee CM (2003) Brain abscess: clinical analysis of 53 cases. J Microbiol Immunol Infect 36:129–136

Khan MA, Greig JR, Jayamohan J (2000) Community-acquired methicillin-resistant *Staphylococcus aureus* brain abscess in an immunocompetent individual. Scand J Infect Dis 32:423–424

Kielian T (2004) Immunopathogenesis of brain abscess. J Neuroinflammation 1:16

Kielian T, Hickey WF (2000) Proinflammatory cytokine, chemokine, and cellular adhesion molecule expression during the acute phase of experimental brain abscess development. Am J Pathol 157:647–658

Kielian T, Barry B, Hickey WF (2001a) CXC chemokine receptor-2 ligands are required for neutrophil-mediated host defense in experimental brain abscesses. J Immunol 166: 4634–4643

Kielian T, Cheung A, Hickey WF (2001b) Diminished virulence of an alpha-toxin mutant of *Staphylococcus aureus* in experimental brain abscesses. Infect Immun 69:6902–6911

Kielian T, Mayes, P, Kielian M (2002) Characterization of microglial responses to *Staphylococcus aureus*: effects on cytokine, costimulatory molecule, and Toll-like receptor expression. J Neuroimmunol 130(1–2):86–99

Kielian T, McMahon M, Bearden ED, Baldwin AC, Drew PD, Esen N (2004a) *S. aureus*-dependent microglial activation is selectively attenuated by the cyclopentenone prostaglandin 15-deoxy-Delta12,14- prostaglandin J2 (15d-PGJ2). J Neurochem 90:1163–1172

Kielian T, Bearden ED, Baldwin AC, Esen N (2004b) IL-1 and TNF-alpha play a pivotal role in the host immune response in a mouse model of *Staphylococcus aureus*-induced experimental brain abscess. J Neuropathol Exp Neurol 63:381–396

Kielian T, Esen N, Bearden ED (2005a) Toll-like receptor 2 (TLR2) is pivotal for recognition of *S. aureus* peptidoglycan but not intact bacteria by microglia. Glia 49:567–576

Kielian T, Haney A, Mayes PM, Garg S, Esen N (2005b) Toll-like receptor 2 modulates the proinflammatory milieu in *Staphylococcus aureus*-induced brain abscess. Infect Immun 73:7428–7435

Kielian T, Phulwani NK, Esen N, Syed MM, Haney AC, McCastlain K, Johnson J (2007) MyD88-dependent signals are essential for the host immune response in experimental brain abscess. J Immunol 178:4528–4537

Koedel U, Angele B, Rupprecht T, Wagner H, Roggenkamp A, Pfister HW, Kirschning CJ (2003) Toll-like receptor 2 participates in mediation of immune response in experimental pneumococcal meningitis. J Immunol 170:438–444

Krcmery V, Fedor-Freybergh PG (2007) Neuroinfections in developed versus developing countries. Neuro Endocrinol Lett 28(Suppl 2):5–6

Kumar K (2000) Neurological complications of congenital heart disease. Indian J Pediatr 67:S15–S19

Laflamme N, Rivest S (2001). Toll-like receptor 4: the missing link of the cerebral innate immune response triggered by circulating gram-negative bacterial cell wall components. Faseb J 15:155–163

Lehnardt S, Lachance C, Patrizi S, Lefebvre S, Follett PL, Jensen FE, Rosenberg PA, Volpe JJ, Vartanian T (2002) The toll-like receptor TLR4 is necessary for lipopolysaccharide-induced oligodendrocyte injury in the CNS. J Neurosci 22:2478–2486

Lehnardt S, Massillon L, Follett P, Jensen FE, Ratan R, Rosenberg PS, Volpe JJ, Vartanian T (2003) Activation of innate immunity in the CNS triggers neurodegeneration through a Toll-like receptor 4-dependent pathway. Proc Natl Acad Sci U S A 100:8514–8519

Maertzdorf J, Osterhaus AD, Verjans GM (2002) IL-17 expression in human herpetic stromal keratitis: modulatory effects on chemokine production by corneal fibroblasts. J Immunol 169:5897–5903

Mathisen GE, Johnson JP (1997) Brain abscess. Clin Infect Dis 25:763–779; quiz 780–761

McClelland CJ, Craig BF, Crockard HA (1978) Brain abscesses in Northern Ireland: a 30 year community review. J Neurol Neurosurg Psychiatry 41:1043–1047

McGeer PL, McGeer EG (2002) Local neuroinflammation and the progression of Alzheimer's disease. J Neurovirol 8:529–538

Menendez Iglesias B, Cerase J, Ceracchini C, Levi G, Aloisi F (1997) Analysis of B7–1 and B7–2 costimulatory ligands in cultured mouse microglia: upregulation by interferon-gamma and lipopolysaccharide and downregulation by interleukin-10, prostaglandin E2 and cyclic AMP-elevating agents. J Neuroimmunol 72:83–93

Nichols JR, Aldrich AL, Mariani MM, Vidlak D, Esen N, Kielian T (2009) TLR2 deficiency leads to increased Th17 infiltrates in experimental brain abscesses. J Immunol 182:7119–7130

Nicolosi A, Hauser WA, Musicco M, Kurland LT (1991) Incidence and prognosis of brain abscess in a defined population: Olmsted County, Minnesota, 1935–1981. Neuroepidemiology 10:122–131

Njambi S, Huttova M, Kovac M, Freybergh PF, Bauer F, Muli JM (2007) Fungal neuroinfections: rare disease but unacceptably high mortality. Neuro Endocrinol Lett 28 Suppl 2:25–26

O'Keefe GM, Nguyen VT, Ping Tang LL, Benveniste EN (2001) IFN-gamma regulation of class II transactivator promoter IV in macrophages and microglia: involvement of the suppressors of cytokine signaling-1 protein. J Immunol 166:2260–2269

Olson JK, Miller SD (2004) Microglia initiate central nervous system innate and adaptive immune responses through multiple TLRs. J Immunol 173:3916–3924

Pasare C, Medzhitov R (2005) Toll-like receptors: linking innate and adaptive immunity. Adv Exp Med Biol 560:11–18

Peiser L, Mukhopadhyay S, Gordon S (2002) Scavenger receptors in innate immunity. Curr Opin Immunol 14:123–128

Persidsky Y, Ghorpade A, Rasmussen J, Limoges J, Liu XJ, Stins M, Fiala M, Way D, Kim KS, Witte MH, Weinand M, Carhart L, Gendelman HE (1999) Microglial and astrocyte chemokines regulate monocyte migration through the blood–brain barrier in human immunodeficiency virus-1 encephalitis. Am J Pathol 155:1599–1611

Phulwani NK, Esen N, Syed MM, Kielian T (2008) TLR2 expression in astrocytes is induced by TNF-alpha- and NF-kappaB-dependent pathways. J Immunol 181:3841–3849

Prasad KN, Mishra AM, Gupta D, Husain N, Husain M, Gupta RK (2006) Analysis of microbial etiology and mortality in patients with brain abscess. J Infect 53:221–227

Rasley A, Anguita J, Marriott I (2002a) *Borrelia burgdorferi* induces inflammatory mediator production by murine microglia. J Neuroimmunol 130:22–31

Rivest S (2003) Molecular insights on the cerebral innate immune system. Brain Behav Immun 17:13–19

Roche M, Humphreys H, Smyth E, Phillips J, Cunney R, McNamara E, O'Brien D, McArdle O (2003) A twelve-year review of central nervous system bacterial abscesses; presentation and aetiology. Clin Microbiol Infect 9:803–809

Ruddy MJ, Shen F, Smith JB, Sharma A, Gaffen SL (2004a) Interleukin-17 regulates expression of the CXC chemokine LIX/CXCL5 in osteoblasts: implications for inflammation and neutrophil recruitment. J Leukoc Biol 76:135–144

Ruddy MJ, Wong GC, Liu XK, Yamamoto H, Kasayama S, Kirkwood KL, Gaffen SL (2004b) Functional cooperation between interleukin-17 and tumor necrosis factor-alpha is mediated by CCAAT/enhancer-binding protein family members. J Biol Chem 279:2559–2567

Schliamser SE, Backman K, Norrby SR (1988) Intracranial abscesses in adults: an analysis of 54 consecutive cases. Scand J Infect Dis 20:1–9

Sell B, Evans J, Horn D (2008) Brain abscess and hereditary hemorrhagic telangiectasia. South Med J 101:618–625

Smith NP, Nelson MR, Moore D, Gazzard BG (1997) Cerebral abscesses in a patient with AIDS caused by methicillin-resistant *Staphylococcus aureus* (MRSA). Int J STD AIDS 8:459–460

Stenzel W, Soltek S, Schluter D, Deckert M (2004) The intermediate filament GFAP is important for the control of experimental murine *Staphylococcus aureus*-induced brain abscess and toxoplasma encephalitis. J Neuropathol Exp Neurol 63:631–640

Stenzel W, Soltek S, Sanchez-Ruiz M, Akira S, Miletic H, Schluter D, Deckert M (2008) Both TLR2 and TLR4 are required for the effective immune response in *Staphylococcus aureus*-induced experimental murine brain abscess. Am J Pathol 172:132–145

Syed MM, Phulwani NK, Kielian T (2007) Tumor necrosis factor-alpha (TNF-alpha) regulates Toll-like receptor 2 (TLR2) expression in microglia. J Neurochem 103:1461–1471

Takeuchi O, Hoshino K, Akira S (2000) Cutting edge: TLR2-deficient and MyD88-deficient mice are highly susceptible to *Staphylococcus aureus* infection. J Immunol 165:5392–5396

Takeuchi O, Sato S, Horiuchi T, Hoshino K, Takeda K, Dong Z, Modlin RL, Akira S (2002) Cutting edge: role of toll-like receptor 1 in mediating immune response to microbial lipoproteins. J Immunol 169:10–14
Tenney JH (1986) Bacterial infections of the central nervous system in neurosurgery. Neurol Clin 4:91–114
Townsend GC, Scheld WM (1998) Infections of the central nervous system. Adv Intern Med 43:403–447
Underhill DM, Gantner B (2004) Integration of Toll-like receptor and phagocytic signaling for tailored immunity. Microbes Infect 6:1368–1373
van der Flier M, Geelen SP, Kimpen JL, Hoepelman IM, Tuomanen EI (2003) Reprogramming the host response in bacterial meningitis: how best to improve outcome? Clin Microbiol Rev 16:415–429
von Herrath MG, Fujinami RS, Whitton JL (2003) Microorganisms and autoimmunity: making the barren field fertile? Nat Rev Microbiol 1:151–157
Weber JR, Moreillon P, Tuomanen EI (2003) Innate sensors for Gram-positive bacteria. Curr Opin Immunol 15:408–415
Winn HR, Mendes M, Moore P, Wheeler C, Rodeheaver G (1979) Production of experimental brain abscess in the rat. J Neurosurg 51:685–690
Witowski J, Pawlaczyk K, Breborowicz A, Scheuren A, Kuzlan-Pawlaczyk M, Wisniewska J, Polubinska A, Friess H, Gahl GM, Frei U, Jorres A (2000) IL-17 stimulates intraperitoneal neutrophil infiltration through the release of GRO alpha chemokine from mesothelial cells. J Immunol 165:5814–5821
Xu J, Ling EA (1995) Induction of major histocompatibility complex class II antigen on amoeboid microglial cells in early postnatal rats following intraperitoneal injections of lipopolysaccharide or interferon-gamma. Neurosci Lett 189:97–100
Ye P, Rodriguez FH, Kanaly S, Stocking KL, Schurr J, Schwarzenberger P, Oliver P, Huang W, Zhang P, Zhang J, Shellito JE, Bagby GJ, Nelson S, Charrier K, Peschon JJ, Kolls JK (2001) Requirement of interleukin 17 receptor signaling for lung CXC chemokine and granulocyte colony-stimulating factor expression, neutrophil recruitment, and host defense. J Exp Med 194:519–527

Toll-Like Receptors in CNS Viral Infections

Hyeon-Sook Suh, Celia F. Brosnan, and Sunhee C. Lee

Contents

Abstract Protection against viral infections is critically dependent upon the early production of significant levels of type 1 interferons and the expression of interferon-stimulated genes that function as the effectors of innate antiviral immunity. Activation of Toll-like receptors on cells of the immune system is known to play an important role in this process. In this chapter we review evidence for a role of TLRs in innate immune responses against viral infections of the central nervous system. By far the most extensive literature pertains to TLR3. Data from various laboratories have shown that TLR3 is expressed in cells endogenous to the CNS, particularly in astrocytes and microglia. Triggering TLR3 by synthetic dsRNA, poly I:C effectively induces innate antiviral responses as well as boosts adaptive immune responses. Additional experiments show cooperative responses between TLRs (3, 7/8 and 9) in mounting an effective antiviral immune response in the periphery.

H.-S. Suh, C.F. Brosnan, and S.C. Lee (✉)
Department of Pathology (Neuropathology), Albert Einstein College of Medicine, 1300 Morris Park Avenue, Bronx, NY, 10461, USA
e-mail: slee@aecom.yu.edu

T. Kielian (ed.), *Toll-like Receptors: Roles in Infection and Neuropathology*, Current Topics in Microbiology and Immunology 336,
DOI 10.1007/978-3-642-00549-7_4,

Perhaps the most exciting data are from patient populations that document the critical role that specific TLRs play in specific CNS infections. Studies also suggest that inappropriate activation of the TLRs can result in a pathogenic outcome rather than a protective one. Since TLR ligands are being actively considered for their antiviral and potential adjuvant effects, this will be an important issue to address in the context of the CNS environment.

Abbreviations

(APOBEC3G)	Apolipoprotein B mRNA-editing enzyme, catalytic polypeptide-like 3G
(*Cig5*: viperin)	Cytomegalovirus-inducible gene 5
CNS	Central nervous system
CpG ODN	Unmethylated CpG oligodeoxynucleotide
dsRNA	Double-stranded RNA
GM-CSF	Granulocyte macrophage colony-stimulating factor
HAART	Highly active antiretroviral therapy
HCMV	Human cytomegalovirus
HIV	Human immunodeficiency virus
IDO	Indoleamine 2,3-dioxygenase
IFN	Interferon
IKK	IκB-activating kinase
IKKi	Inducible IKK
IRF	Interferon regulator factor
iNOS	Inducible Nitric Oxide Synthase
IPS-1	IFN-beta promoter stimulator 1
IRAK	Interlukin-1 receptor-associated Kinase
ISG	Interferon-stimulated gene
ISRE	Interferon-stimulated response elemont
LPS	Lipopolysaccharide
LRR	Leucine-rich repeat
MAPK	Mitogen-activated protein kinase
MCMV	Murine cytomegalovirus
MDA-5	Melanoma differentiation-associated gene 5
MS	Multiple sclerosis
MyD88	Myeloid differentiation factor 88
OAS	2′,5′-Oligoadenylate synthase
PAMP	Pathogen-associated molecular patterns
PCR	Polymerase chain reaction
PKR	dsRNA-dependent protein kinase
Poly I:C	Polyinosinic:polycytidylic acid
RIG-I	Retinoic acid-inducible gene I
RIP1	Receptor-interacting protein 1

RLH	RIG-I-like helicases
STAT1	Signal transducer and activator 1
TANK	TRAF family member-associated NF-κB activator-binding kinase 1
TBK1	TANK-binding kinase 1
TIR	Toll-like receptor IL-1 receptor domain
TLR	Toll-like receptor
TNF-α	Tumor necrosis factor alpha
TRAF	TNFα-receptor associated factor
TRIF	TIR domain-containing adaptor-inducing IFNβ

1 Toll-Like Receptors (TLRs) That Are Viral Sensors

In this chapter we will address the role of TLR as sensors and initiators of the innate immune response to viral infection in the central nervous system (CNS). At least four TLRs (3, 7, 8 and 9) are engaged in viral pathogen-associated molecular patterns (PAMP) recognition (see the introduction to this volume) (Akira et al. 2006). In contrast to other TLRs that recognize components of the cell wall, these TLRs recognize pathogenic nucleic acids and are localized intracellularly, principally in endosomal compartments. TLR3 recognizes viral dsRNA, TLR7 and 8 recognize viral ssRNA, and TLR9 recognizes CpG motifs present in bacterial and viral DNA. Consistent with the endosomal localization, acidic pH is crucial to PAMP recognition and signaling by these TLRs.

The resting CNS is an immune-privileged environment devoid of most immune processes, and the innate immune mechanisms mediating the recognition of CNS infections are largely unknown. Analysis of TLR (1–9) gene expression in the resting CNS detected high expression of TLR3 and TLR9, with the levels of TLR3 transcript reaching that found in the spleen (McKimmie et al. 2005a). The high TLR3 expression has been suggested to reflect the selective pressure RNA viruses have placed on the CNS, as encephalitides caused by RNA viruses are common (see below). The TLR expression levels in the brain have also been found to differ between Balb/c and 129 (derivative of C57Bl/6) strains of mice, which might explain the differential susceptibility of the mouse strains to different microbial organisms (McKimmie et al. 2005a). For example, in C576Bl/6 mice, knockout of TLR3 and particularly TLR9 renders them as susceptible to cytomegalovirus-associated morbidity as the susceptible Balb/c strain (Tabeta et al. 2004).

The expression of TLRs in human CNS has been studied by immunohistochemistry (Bsibsi et al. 2002). While TLR3 and TLR4 immunoreactivity was almost absent from control brains, elevated expression of TLR3 and TLR4 was detected in multiple sclerosis (MS) lesions. Numbers of TLR-expressing cells were strongly increased, with relatively high expression observed, particularly in perivascular areas. Resident microglia and reactive astrocytes were among the cells that expressed TLR3 and TLR4 (Bsibsi et al. 2002).

Primary microglia isolated from adult rhesus monkeys have been examined for TLR transcript expression and response to TLR ligands (Zuiderwijk-Sick et al. 2007). When GM-CSF and M-CSF-treated rhesus microglia were compared, they did not differ in their APC molecule expression and function, but they did differ in their TLR8 expression and response to the TLR8 ligand. Inflammatory cytokine (TNFα and IL-12) production in response to TLR8 ligand was significantly higher in GM-CSF-treated microglia, suggesting that endogenous microglial growth factors can alter their innate responses to infectious stimuli. The remarkable sensitivity of microglia to TLR8-mediated signaling indicated that the CNS is well equipped to detect ssRNA viruses. In addition, it has been reported that TLR8 signaling in neurons acts as a negative regulator of neurite outgrowth and inducer of neuronal apoptosis, indicating the presence of CNS endogenous ligand for TLR8 (Ma et al. 2006).

In vitro, microglia, the resident macrophages of the CNS, express a number of TLRs. These include TLR1, 2, 3, 4, 5, 6, 7, 8, and 9 (examined by real-time PCR in most studies, with the results varying slightly depending on the study and the absence of TLR8 in the mouse). In astrocytes, the expression of TLR is more limited, with most studies reporting the expression of TLR2, TLR3 and TLR4 transcripts (again with some variation depending on the study) (Bsibsi et al. 2002; Carpentier et al. 2005; Farina et al. 2005; Jack et al. 2005; McKimmie and Fazakerley 2005). TLR3 expression has been shown to be particularly prominent in astrocytes regardless of the species or age of the brains from which the cells were isolated. Following ligand activation, CNS cells expressing these receptors release factors that are important not only in the antiviral response but also in the initiation of inflammation and the formation of an adaptive immune response, and perhaps also in repair.

2 Toll-Like Receptor 3 (TLR3)

The vast majority of the work on TLR and CNS viral infections relates to the activation of TLR3. The only known natural ligand for TLR3 is dsRNA, which is found in some RNA viruses such as reovirus. However, it has also been proposed that activation of TLR3 may follow the production of dsRNA intermediates that form transiently during the replication cycle of many of RNA and DNA viruses, suggesting that TLR3 may function as a universal receptor for activation of the innate immune response to viral infections (Schroder and Bowie 2005). It is also possible that TLR3 may be activated by cellular RNA that is released as a result of tissue destruction caused by viral infections of the CNS. In most experimental paradigms, the synthetic dsRNA mimic polyriboinosinic:ribocytidylic acid (poly I:C) has been used as the ligand for TLR3. As noted above, TLR3 is the most abundantly expressed TLR in resting astrocytes, in contrast to resting microglia, which express TLR3 at considerably lower levels (Farina et al. 2005; Olson and Miller 2004; Jack et al. 2005).

2.1 TLR3 Signaling

Following activation with poly I:C, both human and mouse astrocytes express a wide range of cytokines and chemokines characteristic of an innate immune response, with a marked bias towards genes characterized for their role in antiviral responses (Rivieccio et al. 2006; Park et al. 2006). Unlike other TLRs, ligand binding to TLR3 activates a MyD88-*in*dependent pathway of gene induction. The alternative adaptor used in the TLR3 signaling is Toll/IL-1R (TIR) domain-containing adaptor-inducing IFNβ (TRIF), which associates with TLR3 through the TIR-binding domain (Fig. 1) (Akira et al. 2006). It is believed that TRAF6 and receptor-interacting protein-1 (RIP1) are involved in TRIF-dependent NF-κB activation through activation of the canonical IKKs (α, β, and γ), while IKKi (also called IKKε) and possibly TBK1 have recently been shown to be the long-sought interferon regulatory factor 3 (IRF3) kinases. IRF3 is the crucial transcription factor involved in IFNβ gene expression together with NF-κB and MAP kinases, and the induction of the IFNβ gene by TLR3 (and TLR4, as well as IL-1β, see below) is an

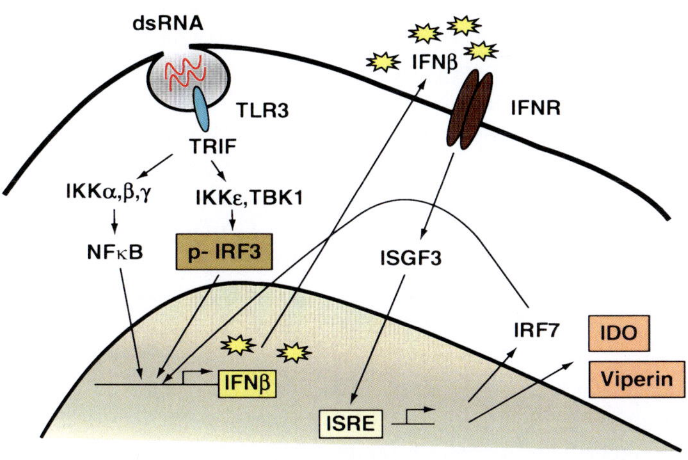

Fig. 1 TLR3 signaling leading to antiviral immunity in astrocytes and microglia. TLR3 located in the intracellular endosomal membrane is triggered by binding with dsRNA such as poly I:C. TLR3, unlike other TLRs, uses the adaptor protein TRIF to transduce signals to activate IRF3 kinases, such as IKKε or TBK1. Both NF-κB and IRF3 are necessary to transactivate the human IFNβ gene (primary response gene). Secreted IFNβ then binds to the type I IFN receptor (IFNR) in an autocrine manner and transactivates the ISGs by forming the ISG factor 3 (ISGF3: STAT1, STAT2, and IRF9) that simulates the IFN-stimulated response element (ISRE) of the gene promoter. One of the genes induced in this manner is the transcription factor IRF7, which synergizes with IRF3 in the activation of IFNβ gene and is necessary for the activation of all human IFNα genes, thereby establishing a positive feedback mechanism for ISG expression. Another positive feedback mechanism appears to include upregulation of the receptor TLR3 itself by poly I:C and other inflammatory stimuli. We have identified IDO and viperin as antiviral effector molecules induced by TLR3, in addition to classical IFN-induced antiviral proteins such as PKR and 2′,5′-oligoadenylate synthetase. (Reproduced with permission from Suh et al. 2007)

important feature that contributes to the antiviral gene (interferon-stimulated genes: ISGs) induction by TLR3. Importantly, the cytosolic viral dsRNA sensors (RIG-I-like helicases: RLH) also activate IRF3 and subsequent IFNβ gene induction through an alternative adaptor, IFNβ promoter stimulator-1 (IPS-1; see below).

The immediate activation of IFNβ (a primary response gene, peaks at ~3 h) through these receptors (TLR3, TLR4, IL-1 receptor and RLH) sets off a second wave of STAT1-dependent ISG induction (secondary response) including IRF7 (Levy et al. 2002). IRF7 synergizes with IRF3 in the induction of type I IFNs (IFNβ and IFNα_{1-12} in humans), type III IFNs (IFN lambda), and certain chemokines (IP-10/CXCL10 and RANTES/CCL5). This aspect of IFNβ-mediated autocrine and paracrine amplification of antiviral genes constitutes a main mechanism for the potent innate immunity elicited by TLR3 and others. Another aspect that is unique to TLR3 (but not other receptors inducing IFNβ) is the positive feedback response built in to TLR3 receptor expression. TLR3 expression has been shown to be uniquely upregulated by many stimuli (Farina et al. 2005; Jack et al. 2005; Bsibsi et al. 2006; Suh et al. 2007). These include the vast majority of proinflammatory cytokines, some regulatory cytokines, TLR ligands including dsRNA (poly I:C) itself, as well as other noxious stimuli such as reactive oxygen species. Furthermore, repeated exposure to the upregulating stimuli appears to result in escalating amounts of TLR3 expression. For example, in human fetal astrocytes, daily exposure to IL-1β or poly I:C for several days not only increases the amounts of TLR transcripts, but also results in increased signaling, which translates to higher ISG (and other inflammatory gene) expression (Suh et al. 2007; Rivieccio et al. 2006). In contrast, repeated exposure to IL-1β reduces the amount of IL-1β-induced gene expression, most likely through receptor downregulation. Similarly, TLR4 expression is not subject to amplification after stimulation with its own ligand or other inflammatory stimuli, and receptor expression is often reduced as a result (Jack et al. 2005). In vivo, infection with neuroinvasive Semliki Forest virus resulted in productive CNS infection followed by rapid upregulation of TLR3 and TLR9 gene expression, which was dependent upon the type-I interferon response (McKimmie et al. 2005b). Similarly, rabies virus infection also increased transcripts for multiple TLRs including TLR3. Together, these features of TLR3 signaling uniquely qualify it as a potent receptor that can elicit significant innate antiviral immune responses or proinflammatory responses that can set off inappropriate activation of the CNS immune system, potentially resulting in reversible and irreversible neural dysfunction (see below).

In contrast to TLR3, other viral sensors (TLR7, 8 and 9) utilize MyD88 as the adaptor to interact with the TIR domain of the receptor, and interaction with the downstream molecules including IRAK4 and TRAF6 ultimately leads to the activation of IKKα, β, and γ and MAP kinases (see the introduction to this volume) (Akira et al. 2006; Kawai and Akira 2007). IRF7 is integral to TLR7 and 8 signaling, as in TLR3. The relevance of these receptors to CNS antiviral immune response is less clear, though TLR7/8 appear to be involved in the antiviral defense in the periphery, mostly in dendritic cells (DCs: conventional and plasmacytoid DCs, see below).

2.2 TLR3 Versus Cytosolic dsRNA Sensors

Recently, an alternative class of cytosolic viral dsRNA sensors termed RIG-I-like helicases (RLH) has been identified. Members of this family include retinoic acid-inducible gene I (RIG-I) and melanoma differentiation-associated gene 5 (MDA-5) (Sen and Sarkar 2005; Akira et al. 2006; Hiscott et al. 2006). While TLR3 signals through the adaptor TRIF, RLHs signal through the adaptor IFN-β promoter stimulator 1 (IPS-1). The relative contributions of these dsRNA sensors to viral recognition and innate immunity are incompletely understood. Through gene targeting experiments in the mouse, Kato et al. have shown that RIG-I is essential for the induction of type I IFNs in fibroblasts and conventional DCs infected with RNA viruses. In contrast, plasmacytoid DCs (also called IFN-producing cells) use the TLR system rather than RIG-I for viral detection (Kato et al. 2005). RNA viruses are also differentially recognized by the two helicases RIG-I and MDA5: while RIG-I is essential for the production of IFNs in response to RNA viruses including paramyxoviruses, influenza virus and Japanese encephalitis virus, MDA5 is critical to the recognition of picornavirus (Kato et al. 2006).

2.3 TLR3 and CNS Glial Cells

In both astrocytes and microglia, the synthetic TLR3 ligand poly I:C has been used to study the function of TLR3. In several studies including our own, poly I:C has been shown to activate a wide array of genes, including cytokines, chemokines, adhesion molecules, co-stimulatory molecules, growth factors, and interferon-stimulated genes (ISGs) (Rivieccio et al. 2006; Jack et al. 2005; Carpentier et al. 2008; Bsibsi et al. 2006). The role of TLR3 (and RLH) as the viral and synthetic dsRNA sensor and in gene activation in the CNS has been examined employing various approaches. These include: (1) the use of TLR3 gene targeting in the mouse; (2) delivery of poly I:C in vitro (external application preferentially activates TLR3, whereas transfection by various methods facilitates stimulation of RLH); (3) TLR3-specific RNA interference; and (4) the use of lysosomal inhibitors that prevent cell acidification (ammonium chloride, for example) by inhibiting signaling via intracellular TLRs (TLR3, 7/8, and 9), which does not affect signaling via RLH. The use of neutralizing antibodies against TLR3 has proven to be ineffective by most laboratories, including our own.

Following stimulation with poly I:C, primary cultures of murine microglia secrete cytokines such as IFNβ, TNFα and IL-6, upregulate TLR3 expression, and show morphological evidence of activation. In microglia isolated from TLR3−/− mice, all of these effects were greatly diminished (Town et al. 2006). Phosphorylation of MAP kinases was also examined as a surrogate marker of dsRNA receptor signaling. This occurred in a time-dependent fashion following poly I:C treatment of wild-type microglia, but showed delayed kinetics in TLR3−/− microglia.

In vivo, microgliosis was clearly evident in wild-type mice following intraventricular injection of poly I:C, but was virtually absent in TLR3−/− mice (Town et al. 2006). These data clearly demonstrate that poly I:C acts primarily through TLR3 in murine microglia both in vitro and in vivo.

In human fetal microglia and astrocytes, we have also shown the role of TLR3 in dsRNA (poly I:C) signaling through the use of TLR3 RNA interference or following lysosomotropic poly I:C agents that inhibit cell acidification (Rivieccio et al. 2006; Suh et al. 2007, 2009; Lee et al. 2006). Taken together, the results demonstrate that externally applied poly I:C is recognized primarily through TLR3. In addition to TLR3, poly I:C has been shown to signal through the cytosolic receptor RLH (MDA5, specifically) (Kato et al. 2006).

In human microglia, gene profiling studies have shown that the types and the magnitude of gene activation following poly I:C stimulation are similar to that following TLR4 stimulation by LPS (Suh et al. 2009; Lee et al. 2006). Furthermore, western blot comparison of cell signaling between poly I:C-activated and LPS-activated human microglia demonstrate a remarkably similar pattern in terms of the kinetics and levels of transcription factor/second messenger induction and phosphorylation belonging to several key signaling pathways (NF-κB, IRF3, MAP kinases and PI3K pathway). This translates into the production of antiviral proteins in similar quantities and with similar kinetics by poly I:C and LPS, which also translates into equally potent antiviral activities of poly I:C and LPS in primary human microglia in culture when studied in the context of HIV replication. These data confirm that in microglia, in addition to the MyD88-dependent pathway (Esen and Kielian 2006), TLR4 can activate the MyD88-independent (TRIF-dependent) pathway of IRF3 activation, resulting in IFNβ production, akin to TLR3 signaling, resulting in effective innate antiviral immunity.

In human fetal astrocytes, activation of TLR3 by poly I:C also activates >1,000 genes ranging from cytokines to chemokines, ISGs, adhesion molecules and growth factors (Rivieccio et al. 2006). As predicted, IFNβ gene induction occurred at around 3 h, followed by delayed STAT1 phosphorylation (a surrogate marker of type I IFN production). The induction of many ISGs including viperin and indoleamine 2,3-dioxygenase (IDO, see below) also translated into effective innate antiviral immunity against HIV and human cytomegalovirus (HCMV) in astrocytes (Rivieccio et al. 2006; Suh et al. 2007). In astrocytes, however, TLR4 signaling is minimally activated by LPS, in part due to the absence of the surface LPS-binding protein, CD14. In contrast to the lack of response to LPS, astrocyte stimulation with IL-1 activates >1,000 genes in astrocytes (John et al. 2005). The types of genes induced, including cytokine, chemokine and antiviral genes, are similar to those elicited by TLR3 activation. In addition, we found that IL-1 activates IRF3-dependent IFNβ gene induction in astrocytes, akin to TLR4 signaling in microglia (Rivieccio et al. 2005). However, in contrast to microglia stimulated with LPS or poly I:C-stimulated astrocytes, IL-1-stimulated astrocytes show much lower amounts of antiviral gene expression when examined by microarray as well as by quantitative real-time PCR (Rivieccio et al. 2006; Suh et al. 2007). The latter may explain why the antiviral immunity elicited by IL-1 is ineffective compared with poly I:C in astrocytes (Suh et al. 2007; Rivieccio et al. 2006).

Poly I:C-induced gene expression has also been compared to LPS-induced gene expression in cultures of adult human astrocytes derived postmortem (Meeuwsen et al. 2003; Bsibsi et al. 2006). Poly I:C induced a wide array of gene expression in adult astrocytes, with several growth factor genes being among the most highly induced genes. LPS induced a much more limited response in astrocytes and lacked the ability to activate many of the potentially neuroprotective mediators and anti-inflammatory cytokines induced by poly I:C. Poly I:C-conditioned medium, but not LPS-conditioned medium, promoted the survival of neurons in organotypic human brain slice cultures, implying that TLR3 ligation (by either poly I:C or viral dsRNA) in human brain may elicit neuroprotective gene expression in astrocytes (Bsibsi et al. 2006).

2.4 Mechanisms of Antiviral Immunity Elicited by TLR3 In Vitro

Poly I:C-induced antiviral gene expression in astrocytes and microglia translates into antiviral activity. This functional confirmation is important because IL-1 also induces IFNβ and ISGs in astrocytes but does not display antiviral activity when tested against HIV and HCMV. Adenovirus-mediated overexpression of a dominant-negative IRF3 demonstrated that the antiviral activity of poly I:C was dependent on IRF3, confirming the pivotal role that this transcription factor plays in the innate antiviral immune response (Suh et al. 2007).

Our recent work on human astrocytes has elucidated the roles of indoleamine 2,3-dioxygenase (IDO) and viperin in the innate host defense mechanism against HIV (Suh et al. 2007; Rivieccio et al. 2006). We found that while IDO knockdown significantly reduced antiviral activity, the extent to which IDO individually contributed to the overall antiviral response was small. This was not surprising, since poly I:C induces/activates other well-known antiviral proteins such as dsRNA-dependent protein kinase (PKR) and 2′,3′-oligoadenylate synthase. We also found that poly I:C induces viperin/Cig5, another protein with known antiviral activity against HCMV and HIV (Chin and Cresswell 2001; Rivieccio et al. 2006). Novel anti-HIV proteins that are inducible by poly I:C and interferons include the recently discovered apolipoprotein B mRNA-editing enzyme catalytic polypeptide-like 3G (APOBEC3G), which was isolated as an anti-HIV protein suppressed by the viral Vif protein (Huthoff and Malim 2005). In human astrocytes and microglia, APOBEC3G is robustly induced by poly I:C in an IRF3-dependent manner (Suh et al. 2009). Together, these studies add IDO, viperin and APOBEC3G to the long list of newly discovered anti-HIV proteins that are induced by TLR3 ligation in human glial cells.

It is important to recognize that data obtained in mouse macrophage/microglia studies may not be directly applicable to the human condition. For example, the reported proapoptotic activity of TLR (IRF3- and IFNβ-dependent) in murine macrophages is apparently related to IFNβ's effect on the inducible nitric oxide synthase (iNOS) promoter, which leads to an increase in the production of nitric oxide (NO) (Jung et al. 2005). Since iNOS expression under most circumstances is negligible in human macrophages and microglia, some of the reported toxic properties of TLR ligands may not occur in human microglia. In fact, we do not see increased apoptosis of microglia when they are exposed to either TLR3 or TLR4 ligands. Furthermore, we have shown that the expression of the rate-limiting enzyme in tryptophan catabolism, IDO, which also has antiviral activity against several neurotropic viruses, is induced by poly I:C in microglia and astrocytes (Suh et al. 2007). Importantly, the enzyme activity of IDO (which correlates with its antiviral activity) is abolished in the presence of NO, without concurrent suppression of IDO protein expression. This may explain the species-dependent activity of macrophage antimicrobial proteins, in that IDO may play a more significant role in human macrophages (and microglia) than in rodent macrophages, since human and rodent macrophages differ in terms of iNOS expression (Daubener and MacKenzie 1999; Hucke et al. 2004). IDO expression is regulated at multiple steps by transcriptional, translational, and posttranslational mechanisms. In several cell systems, IDO expression and activity have been shown to be inhibited by NO. For example, NO enhances proteasomal degradation of IDO, and NO donors inactivate IDO by inducing nitration of critical tyrosine residues (Fujigaki et al. 2006). During our investigation of the relationship between IDO and iNOS in human astrocytes, we first observed that IL-1 induces iNOS expression but very little IDO expression, while poly I:C induces IDO expression with no iNOS expression (Suh et al. 2007). This was surprising because poly I:C induces robust amounts of iNOS (and NO) in rodent astrocytes. We also found that IDO protein expression and IDO activity could be completely dissociated. While IL-1 increased the amount of IDO protein expression induced by IFNγ, IL-1 inactivated IDO enzyme activity by inducing iNOS. IL-1 is an essential cytokine for human iNOS induction in several cell types (astrocytes, chondrocytes and hepatocytes). Thus, IDO function (not limited to antiviral activity) in human cells in vivo could be severely restricted in a cytokine environment where IL-1 is produced.

Because the innate immune response is not pathogen specific, we tested the antiviral activity of poly I:C in human CMV infection of astrocytes in vitro. Human astrocytes infected with HCMV in vitro essentially recapitulate the cytopathic effects of their in vivo counterparts. Poly I:C prevented replication of HCMV as well as the cytopathic effects (multinucleation and cytolysis) in astrocytes. Unlike HIV, HCMV has been reported to activate IRF3 as well as many ISGs in infected cells. The relationships between interferons and HCMV gene induction are complex. For example, IFNγ-activated sequence-like consensus sequences are present in HCMV immediate-early genes, and these sequences are required for HCMV growth at *low viral doses* (Netterwald et al. 2005). Thus, IFNγ could paradoxically increase HCMV replication when the viral concentration is low. Indeed, IFNγ was not effective against HCMV at low viral concentrations, whereas poly I:C continued to suppress HCMV replication under these conditions. These results point to some of the important differences in the antiviral immunity generated by dsRNA and interferons and demonstrate that dsRNA signaling triggers an effective innate antiviral immunity in vitro.

2.5 Antiviral Immunity Elicited by TLR3 In Vivo

The role of TLR3 in murine experimental encephalitides has been examined in TLR3−/− mice (Wang et al. 2004). In mice infected with West Nile Virus, TLR3 plays a critical role in facilitating viral entry into the CNS, resulting in lethal encephalitis. This process was attributable to TNFα-dependent loss of blood–brain barrier function and inflammatory cell infiltration into the CNS. Thus, in this model, a TLR3-mediated increase in cytokine production and inflammation was associated with death even though the viral burden in the periphery was reduced. The results of this study indicate that TLR3-induced inflammation could play a deleterious role in viral pathogenesis, regardless of its effect on innate antiviral immunity.

In another study, the role of TLR3 was examined in wild-type and TLR3−/− mice after infections with lymphocytic choriomeningitis virus, vesicular stomatitis virus (VSV), murine CMV or reovirus, with no differences observed in viral clearance or adaptive immune responses (Edelmann et al. 2004). This study contrasts with another, which reported a significant (although less important than TLR9) role for TLR3 in defense against systemic murine CMV infection (Tabeta et al. 2004). Therefore, the role of TLR3 appears to differ depending on the virus as well as the model system studied.

Multiple factors clearly contribute to the relative importance that TLR3 plays in the pathogenesis of CNS viral infections, and influence whether the outcome will be protective or detrimental. These factors include: (1) whether dsRNA intermediates are produced during the viral replication cycle; (2) whether TLR3 or RLH are used to recognize specific viral dsRNA; and (3) whether viruses have devised a system to counteract the host innate antiviral immune response. For example, West Nile Virus has been shown to interfere with poly I:C-induced activation of IRF3 (Scholle and Mason 2005). The susceptibility of inbred laboratory mice to West Nile Virus (as opposed to wild mice that are resistant) has been linked to a mutation within the 2′,5′-oligoadenylate synthetase gene family (OAS) (Mashimo et al. 2002). This is interesting in light of the recent finding suggesting that a polymorphism in the OAS gene family might play a role in susceptibility to human West Nile Virus infection (Yakub et al. 2005). Furthermore, the hepatitis C virus serine protease NS3/4A is known to cleave IPS-1, an adaptor protein critical to the RLH signaling pathway (Foy et al. 2003). This results in the dislocation of the IPS-1 protein from mitochondria, with the resulting failure to activate IRF3 and NF-κB and to produce IFNβ, the key interferon in innate antiviral immunity.

The most convincing evidence for a role for TLR3 in human antiviral immunity has been provided by the recent study of Zhang et al. (2007). In this report, they show that although redundant antiviral pathways appear to exist in the defense against most microbes, TLR3 signaling is vital to immunity to herpes simplex virus (HSV-1) encephalitis in humans. This study describes a heterozygous TLR3 mutation in two unrelated French children with HSV-1 encephalitis that was detected in the extracellular, ligand-binding domain that contains 23 leucine-rich repeats (LRRs).

The P544S mutation in these children occurred in the LRR20 domain, which is thought to be critical for ligand binding and receptor multimerization. The patients' fibroblasts showed impaired responsiveness to poly I:C in the production of interferons, demonstrating a dominant negative effect of the mutation. In contrast, the patients' fibroblasts responded normally to IL-1 and TNFα. These fibroblasts were also defective in the control of HSV-1 and VSV, and showed enhanced cell death following infection, while their responses to four other viruses were not affected. Interestingly, PBMCs, myeloid DCs, and plasmacytoid DCs, as well as keratinocytes from these patients responded normally to poly I:C, suggesting that it is TLR3 expression in endogenous CNS cells that plays a critical role in the defense against HSV-1 encephalitis.

Additional support for the critical role of TLR3, as well as TLR9, comes from the observation that patients with a point mutation in UNC93B, an endoplasmic reticulum protein required for intracellular TLR (TLR3, 7/8 & 9) signaling, are also susceptible to HSV-1 encephalitis, and the patients' fibroblasts display impaired interferon production in response to poly I:C (Casrouge et al. 2006).

TLR3 upregulation has been observed in vivo in experimental simian immunodeficiency virus (SIV) infection as well as in human brains with HIV encephalitis (Sanghavi and Reinhart 2005; Suh et al. 2007). These results suggest that enhanced TLR3 expression during viral infection could predispose the cells to a greater response. In addition to innate immunity, dsRNA could also boost adaptive immunity against viral infection; for example, by cross-priming viral-specific CD8 + CTLs. These findings together support the notion that TLR3 is upregulated during CNS viral infection, but that effective antiviral TLR3 signaling does not occur due to a lack of available ligands and/or inhibition by the virus. This suggests that TLR3 signaling may provide a potential therapeutic strategy against viral infections such as HIV/AIDS. In human brains with HIV encephalitis, in addition to TLR3, we have found an increase in transcripts for IFNβ and IFN-dependent antiviral proteins such as IDO and viperin (Suh et al. 2007). Significant IFNβ production and activation of innate immunity has also been demonstrated in experimental SIV infection of the brain (Barber et al. 2004). These data suggest that some degree of antiviral innate immune response is activated in vivo during viral encephalitides, although the exact mechanism involved in the induction of antiviral immune molecules is not known. We propose that a chronic suboptimal stimulation of the CNS innate immune response occurs in viral encephalitis via inflammatory cytokines such as IL-1. This may be different from a more effective antiviral immunity that can be triggered through innate immune receptors such as TLR3.

2.6 TLR, Adaptive Immunity and Autoimmune Responses

In vitro ligation of TLR3 on human microglia has been shown to significantly elevate major histocompatibility complex (MHC) antigen and co-stimulatory molecule expression, and to lead to the induction of high levels of IFNα, IL-12p40 and IL-23.

Co-culturing T cells with poly I:C-preactivated microglia led to elevated levels of IFNγ and Th1 polarization. In contrast to TLR3, ligation of TLR2 or TLR4 had smaller or inconsistent effects on MHC II induction, IFNγ production and Th1 polarization (Jack et al. 2007). Similar surface APC antigen upregulation, effector molecule expression and innate cytokine/chemokine expression has also been observed in poly I:C-activated murine astrocytes (Carpentier et al. 2005), in addition to murine microglia (Olson and Miller 2004). Therefore, TLR3 ligation in CNS glia appears to polarize the cytokine environment towards Th1 rather than Th2. While these properties could help boost the innate immune response against certain pathogens and viruses, inappropriate stimulation of TLRs could potentially result in autoimmune destruction of the CNS cells and tissues (Carpentier et al. 2008) (also see below). The Th1 polarization by TLR3 ligation is of particular relevance in terms of the microglial activation phenotypes (M1 vs. M2). Given that microglial activation is a much more nuanced process and that not all activated microglia are damaging to neurons (Schwartz et al. 2006; Gordon 2003), TLR ligands could potentially influence the glial activation phenotypes, thereby influencing neuronal well-being. The action of poly I:C as an adjuvant that enhances adaptive immune responses has also been shown in vivo. The adjuvant effects of poly I:C, as shown by antigen-dependent antibody production and antigen-dependent CD8 + T cell expansion, were shown to involve the cooperative activation of TLR and cytoplasmic RNA helicase pathways (Kumar et al. 2008).

2.7 Effect of Systemic or Intracerebral Poly I:C on Brain Function and Inflammation

Poly I:C has been used to mimic the acute phase of a viral infection. Cunningham et al. have investigated the spectrum of sickness behavioral changes induced by poly I:C in C57BL/6 mice and the CNS expression of inflammatory mediators that may underlie this (Cunningham et al. 2007). Poly I:C dose-dependently induced sickness behavior and changes in body weight and temperature accompanied by elevated plasma TNFα, IL-6 and IFNβ levels. Cytokine synthesis within the CNS was dominated by IL-6. These results demonstrated clear CNS effects of peripheral TLR3 stimulation and are useful for studying aspects of the effects of systemic viral infection on brain function. Interestingly, the authors found that the response to a second administration of poly I:C (1–3 weeks later) was similar to the first, indicating that in vivo behavioral and metabolic tolerance to poly I:C does not occur. This result is reminiscent of the in vitro glial cell response that showed amplified responses (rather than tolerance) with repeated exposure to poly I:C, an effect we attributed to increased TLR3 expression.

As maternal viral infection is known to increase the risk for schizophrenia in the offspring, poly I:C has been used to activate the maternal immune response in order to elicit changes in the developing brain and induce schizophrenia-like behavior in the offspring of mice (Patterson 2007). Apparently, this model of schizophrenia and brain change is also mediated by IL-6. Studies like these demonstrate the double-edged nature of the innate antiviral immune response in that, although it could initially help combat viral infections, chronic inappropriate stimulation of TLR3 might result in adverse effects.

3 Roles for Other TLRs in Virus Infections in the CNS

Very little is known about the functions of other TLRs in the CNS, particularly as pertains to viral infections. However, as noted earlier, encephalitides caused by RNA viruses (polio, rabies, West Nile, and Japanese encephalitis) are common, and either meningitis or encephalitis may also accompany infections caused by DNA viruses or lentiviruses. Thus, it is likely that TLR are activated as part of the innate immune response to these infections in the CNS, and in this regard it may be instructive to consider what has been observed in the periphery.

In chronic untreated HIV infections associated with viremia, multiple TLRs have been found to be increased in the periphery, and this is associated with increased responsiveness to TLR ligands (Lester et al. 2008). Evidence also supports the conclusion that HIV can directly stimulate immune cells via TLRs. In dendritic cells and macrophages, guanosine (G)- and uridine (U)-rich ssRNA oligonucleotides derived from HIV induce the secretion of IFNα and proinflammatory as well as regulatory cytokines (Heil et al. 2004; Beignon et al. 2005). Further studies have shown that recognition of GU-rich ssRNA is species specific and that the TLRs involved in recognizing ssRNA in human are most likely TLR7 and TLR8. These data show that HIV-encoded TLR 7/8 ligands may contribute directly to the immune activation observed during viremic HIV infection (Meier et al. 2007). However, TLR7/8 triggering may modulate HIV infection per se. TLR7/8 stimulation by HIV ssRNA or the synthetic compound R848 has been shown to suppress HIV infection in lymphoid aggregate cultures and PBMC. Unlike lymphoid tissues *acutely* infected with HIV, treatment of *latently* infected promonocytic cells with TLR7/8 ligands induces the release of HIV virions (Schlaepfer et al. 2006). Therefore, TLR7/8 could prevent new infections while promoting viral expression from latent reservoirs. In human microglial cultures, we also find that while poly I:C prevents new productive infection by HIV, it promotes viral expression from latently infected promonocytic cells (U1 cells that contain HIV provirus). Therefore, compounds triggering TLR (3 or 7/8) may be attractive drug candidates for an antireservoir therapy in individuals on HAART (Lehrman et al. 2005).

A role for cooperative effects between different TLRs has been shown in a recent study that examined the activation of a protective response to the DNA virus MCMV in plasmacytoid dendritic cells (Zucchini et al. 2008). Using mice in which the genes for TLR 7 and/or 9 had been inactivated, upon assessing the production of IFNα/β, IL-12 p40 and TNFα, as well as lethality, the data show that TLR9 is required for IL-12p70 production, but that only the combined loss of TLR7 and 9 renders mice as susceptible to MCMV as MyD88–/– mice. This effect was associated with a dramatic decrease in systemic IFNα/β and an increase in viral load. Thus, these data lend further support to the conclusion that DNA viruses may activate TLRs involved in the recognition of viral RNA intermediates, in agreement with recent studies with vesicular stomatitis virus (Waibler et al. 2007).

4 Conclusions and Future Directions

In this chapter we have reviewed the evidence for TLR activation in the CNS, and whether this activation plays a role in providing protection against viral infections. TLRs that would be expected to be activated by viral infections include TLRs 3, 7, 8 and 9. Studies from several groups have clearly shown that the TLR3 ligand poly I:C activates an inflammatory cascade in both microglia and astrocytes, with a strong bias towards the production of proteins with known antiviral properties. These proteins clearly have antiviral activity in vitro, but the roles of specific TLRs in providing protection in vivo remain more controversial when tested in specific TLR knockout mice. However, the roles of specific TLRs may be masked by the need for cooperative effects between different TLRs to achieve potent antiviral activity, and this will clearly be an important area of research for future studies. Nevertheless, a role for TLR3 in protection against herpes encephalitis in humans has now been clearly documented, and in addition confirms that DNA viruses may activate TLRs that are thought to be predominantly responsive to RNA viruses. This is thought to occur through the transient formation of dsRNA intermediates, supporting a role for cross-activation and redundancy within the TLR/RLH pathways involved in the innate immune response to viruses. An additional area of active research is the potential use of TLR ligands as adjuvants for the formation of an effective adaptive immune response. However, data from animal studies has again demonstrated that the protective effects of TLR activation may be difficult to separate from pathogenic effects in the CNS, which is particularly vulnerable to inflammatory events. This will also be an important area to address in future experimentation.

Acknowledgments The authors and the work described in this chapter are in part supported by the PHS grants RO1 MH55477, Molecular Neuropathology Training Grant T32 NS007098, and the Einstein CFAR P30 AI051519, and NMSS RG3827A5. We thank Dr. Mark Rivieccio for his work described in this chapter. We also thank the Einstein Human Fetal Tissue Repository, the National NeuroAIDS Tissue Consortium (Manhattan HIV Brain Bank, Dr. Susan Morgello, director), and the NIH AIDS Research and Reference Reagent Program for tissues and reagents that were essential for generating data described in this chapter. We apologize to those whose relevant work could not be cited due to space limitations.

References

Akira S, Uematsu S, Takeuchi O (2006) Pathogen recognition and innate immunity. Cell 124:783–801

Barber SA, Herbst DS, Bullock BT, Gama L, Clements JE (2004) Innate immune responses and control of acute simian immunodeficiency virus replication in the central nervous system. J Neurovirol 10(Suppl 1):15–20

Beignon AS, McKenna K, Skoberne M, Manches O, DaSilva I, Kavanagh DG, Larsson M, Gorelick RJ, Lifson JD, Bhardwaj N (2005) Endocytosis of HIV-1 activates plasmacytoid dendritic cells via Toll-like receptor-viral RNA interactions. J Clin Invest 115:3265–3275

Bsibsi M, Persoon-Deen C, Verwer RW, Meeuwsen S, Ravid R, van Noort JM (2006) Toll-like receptor 3 on adult human astrocytes triggers production of neuroprotective mediators. Glia 53:688–695

Bsibsi M, Ravid R, Gveric D, van Noort JM (2002) Broad expression of Toll-like receptors in the human central nervous system. J Neuropathol Exp Neurol 61:1013–1021

Carpentier PA, Begolka WS, Olson JK, Elhofy A, Karpus WJ, Miller SD (2005) Differential activation of astrocytes by innate and adaptive immune stimuli. Glia 49:360–374

Carpentier PA, Duncan DS, Miller SD (2008) Glial toll-like receptor signaling in central nervous system infection and autoimmunity. Brain Behav Immun 22:140–147

Casrouge A, Zhang SY, Eidenschenk C, Jouanguy E, Puel A, Yang K, Alcais A, Picard C, Mahfoufi N, Nicolas N, Lorenzo L, Plancoulaine S, Senechal B, Geissmann F, Tabeta K, Hoebe K, Du X, Miller RL, Heron B, Mignot C, de Villemeur TB, Lebon P, Dulac O, Rozenberg F, Beutler B, Tardieu M, Abel L, Casanova JL (2006) Herpes simplex virus encephalitis in human UNC-93B deficiency. Science 314:308–312

Chin KC, Cresswell P (2001) Viperin (cig5), an IFN-inducible antiviral protein directly induced by human cytomegalovirus. Proc Natl Acad Sci USA 98:15125–15130

Cunningham C, Campion S, Teeling J, Felton L, Perry VH (2007) The sickness behaviour and CNS inflammatory mediator profile induced by systemic challenge of mice with synthetic double-stranded RNA (poly I:C). Brain Behav Immun 21:490–502

Daubener W, MacKenzie CR (1999) IFN-gamma activated indoleamine 2,3-dioxygenase activity in human cells is an antiparasitic and an antibacterial effector mechanism. Adv Exp Med Biol 467:517–524

Edelmann KH, Richardson-Burns S, Alexopoulou L, Tyler KL, Flavell RA, Oldstone MB (2004) Does Toll-like receptor 3 play a biological role in virus infections? Virology 322:231–238

Esen N, Kielian T (2006) Central role for MyD88 in the responses of microglia to pathogen-associated molecular patterns. J Immunol 176:6802–6811

Farina C, Krumbholz M, Giese T, Hartmann G, Aloisi F, Meinl E (2005) Preferential expression and function of Toll-like receptor 3 in human astrocytes. J Neuroimmunol 159:12–19

Foy E, Li K, Wang C, Sumpter R Jr, Ikeda M, Lemon SM, Gale M Jr (2003) Regulation of interferon regulatory factor-3 by the hepatitis C virus serine protease. Science 300:1145–1148

Fujigaki H, Saito K, Lin F, Fujigaki S, Takahashi K, Martin BM, Chen CY, Masuda J, Kowalak J, Takikawa O, Seishima M, Markey SP (2006) Nitration and inactivation of IDO by peroxynitrite. J Immunol 176:372–379

Gordon S (2003) Alternative activation of macrophages. Nat Rev Immunol 3:23–35

Heil F, Hemmi H, Hochrein H, Ampenberger F, Kirschning C, Akira S, Lipford G, Wagner H, Bauer S (2004) Species-specific recognition of single-stranded RNA via toll-like receptor 7 and 8. Science 303:1526–1529

Hiscott J, Lin R, Nakhaei P, Paz S (2006) MasterCARD: a priceless link to innate immunity. Trends Mol Med 12:53–56

Hucke C, MacKenzie CR, Adjogble KD, Takikawa O, Daubener W (2004) Nitric oxide-mediated regulation of gamma interferon-induced bacteriostasis: inhibition and degradation of human indoleamine 2,3-dioxygenase. Infect Immun 72:2723–2730

Huthoff H, Malim MH (2005) Cytidine deamination and resistance to retroviral infection: towards a structural understanding of the APOBEC proteins. Virology 334:147–153

Jack CS, Arbour N, Blain M, Meier UC, Prat A, Antel JP (2007) Th1 polarization of CD4 + T cells by Toll-like receptor 3-activated human microglia. J Neuropathol Exp Neurol 66:848–859
Jack CS, Arbour N, Manusow J, Montgrain V, Blain M, McCrea E, Shapiro A, Antel JP (2005) TLR signaling tailors innate immune responses in human microglia and astrocytes. J Immunol 175:4320–4330
John GR, Lee SC, Song X, Rivieccio M, Brosnan CF (2005) IL-1-regulated responses in astrocytes: relevance to injury and recovery. Glia 49:161–176
Jung DY, Lee H, Jung BY, Ock J, Lee MS, Lee WH, Suk K (2005) TLR4, but not TLR2, signals autoregulatory apoptosis of cultured microglia: a critical role of IFN-beta as a decision maker. J Immunol 174:6467–6476
Kato H, Sato S, Yoneyama M, Yamamoto M, Uematsu S, Matsui K, Tsujimura T, Takeda K, Fujita T, Takeuchi O, Akira S (2005) Cell type-specific involvement of RIG-I in antiviral response. Immunity 23:19–28
Kato H, Takeuchi O, Sato S, Yoneyama M, Yamamoto M, Matsui K, Uematsu S, Jung A, Kawai T, Ishii KJ, Yamaguchi O, Otsu K, Tsujimura T, Koh CS, Reis e Sousa, Matsuura Y, Fujita T, Akira S (2006) Differential roles of MDA5 and RIG-I helicases in the recognition of RNA viruses. Nature 441:101–105
Kawai T, Akira S (2007) Antiviral signaling through pattern recognition receptors. J Biochem 141:137–145
Kumar H, Koyama S, Ishii KJ, Kawai T, Akira S (2008) Cutting edge: cooperation of IPS-1- and TRIF-dependent pathways in poly IC-enhanced antibody production and cytotoxic T cell responses. J Immunol 180:683–687
Lee SC, Suh HS, Rivieccio M, Zhao ML, Brosnan CF (2006) TLR3 and antiviral response in astrocytes and microglia (Abstract in 7th International Symposium on NeuroVirology). J Neurovirol vol 12
Lehrman G, Hogue IB, Palmer S, Jennings C, Spina CA, Wiegand A, Landay AL, Coombs RW, Richman DD, Mellors JW, Coffin JM, Bosch RJ, Margolis DM (2005) Depletion of latent HIV-1 infection in vivo: a proof-of-concept study. Lancet 366:549–555
Lester RT, Yao XD, Ball TB, McKinnon LR, Kaul R, Wachihi C, Jaoko W, Plummer FA, Rosenthal KL (2008) Toll-like receptor expression and responsiveness are increased in viraemic HIV-1 infection. AIDS 22:685–694
Levy DE, Marie I, Smith E, Prakash A (2002) Enhancement and diversification of IFN induction by IRF-7-mediated positive feedback. J Interferon Cytokine Res 22:87–93
Ma Y, Li J, Chiu I, Wang Y, Sloane JA, Lu J, Kosaras B, Sidman RL, Volpe JJ, Vartanian T (2006) Toll-like receptor 8 functions as a negative regulator of neurite outgrowth and inducer of neuronal apoptosis. J Cell Biol 175:209–215
Mashimo T, Lucas M, Simon-Chazottes D, Frenkiel MP, Montagutelli X, Ceccaldi PE, Deubel V, Guenet JL, Despres P (2002) A nonsense mutation in the gene encoding 2′-5′-oligoadenylate synthetase/L1 isoform is associated with West Nile virus susceptibility in laboratory mice. Proc Natl Acad Sci USA 99:11311–11316
McKimmie CS, Fazakerley JK (2005) In response to pathogens, glial cells dynamically and differentially regulate Toll-like receptor gene expression. J Neuroimmunol 169:116–125
McKimmie CS, Johnson N, Fooks AR, Fazakerley JK (2005a) Viruses selectively upregulate Toll-like receptors in the central nervous system. Biochem Biophys Res Commun 336:925–933
McKimmie CS, Johnson N, Fooks AR, Fazakerley JK (2005b) Viruses selectively upregulate Toll-like receptors in the central nervous system. Biochem Biophys Res Commun 336:925–933
Meeuwsen S, Persoon-Deen C, Bsibsi M, Ravid R, van Noort JM (2003) Cytokine, chemokine and growth factor gene profiling of cultured human astrocytes after exposure to proinflammatory stimuli. Glia 43:243–253
Meier A, Alter G, Frahm N, Sidhu H, Li B, Bagchi A, Teigen N, Streeck H, Stellbrink HJ, Hellman J, van LJ, Altfeld M (2007) MyD88-dependent immune activation mediated by human immunodeficiency virus type 1-encoded Toll-like receptor ligands. J Virol 81:8180–8191

Netterwald J, Yang S, Wang W, Ghanny S, Cody M, Soteropoulos P, Tian B, Dunn W, Liu F, Zhu H (2005) Two gamma interferon-activated site-like elements in the human cytomegalovirus major immediate-early promoter/enhancer are important for viral replication. J Virol 79:5035–5046

Olson JK, Miller SD (2004) Microglia initiate central nervous system innate and adaptive immune responses through multiple TLRs. J Immunol 173:3916–3924

Park C, Lee S, Cho IH, Lee HK, Kim D, Choi SY, Oh SB, Park K, Kim JS, Lee SJ (2006) TLR3-mediated signal induces proinflammatory cytokine and chemokine gene expression in astrocytes: differential signaling mechanisms of TLR3-induced IP-10 and IL-8 gene expression. Glia 53:248–256

Patterson PH (2007) Neuroscience. Maternal effects on schizophrenia risk. Science 318:576–577

Rivieccio MA, John GR, Song X, Suh HS, Zhao Y, Lee SC, Brosnan CF (2005) The cytokine IL-1beta activates IFN response factor 3 in human fetal astrocytes in culture. J Immunol 174:3719–3726

Rivieccio MA, Suh HS, Zhao Y, Zhao ML, Chin KC, Lee SC, Brosnan CF (2006) TLR3 ligation activates an antiviral response in human fetal astrocytes: a role for viperin/cig5. J Immunol 177:4735–4741

Sanghavi SK, Reinhart TA (2005) Increased expression of TLR3 in lymph nodes during simian immunodeficiency virus infection: implications for inflammation and immunodeficiency. J Immunol 175:5314–5323

Schlaepfer E, Audige A, Joller H, Speck RF (2006) TLR7/8 triggering exerts opposing effects in acute versus latent HIV infection. J Immunol 176:2888–2895

Scholle F, Mason PW (2005) West Nile virus replication interferes with both poly(I:C)-induced interferon gene transcription and response to interferon treatment. Virology 342:77–87

Schroder M, Bowie AG (2005) TLR3 in antiviral immunity: key player or bystander? Trends Immunol 26:462–468

Schwartz M, Butovsky O, Bruck W, Hanisch UK (2006) Microglial phenotype: is the commitment reversible? Trends Neurosci 29:68–74

Sen GC, Sarkar SN (2005) Hitching RIG to action. Nat Immunol 6:1074–1076

Suh HS, Zhao ML, Choi N, Belbin TJ, Brosnan CF, Lee SC: TLR3 and TLR4 are innate antiviral immune receptors in human microglia: role of IRF3 in modulating antiviral and inflammatory response in the CNS. Virology 2009 (In Press)

Suh HS, Zhao ML, Rivieccio M, Choi S, Connolly E, Zhao Y, Takikawa O, Brosnan CF, Lee SC (2007) Astrocyte indoleamine 2,3-dioxygenase (IDO) is induced by the TLR3 ligand poly IC: mechanism of induction and role in anti-viral response. J Virol 81(18):9838–9850

Tabeta K, Georgel P, Janssen E, Du X, Hoebe K, Crozat K, Mudd S, Shamel L, Sovath S, Goode J, Alexopoulou L, Flavell RA, Beutler B (2004) Toll-like receptors 9 and 3 as essential components of innate immune defense against mouse cytomegalovirus infection. Proc Natl Acad Sci USA 101:3516–3521

Town T, Jeng D, Alexopoulou L, Tan J, Flavell RA (2006) Microglia recognize double-stranded RNA via TLR3. J Immunol 176:3804–3812

Waibler Z, Detje CN, Bell JC, Kalinke U (2007) Matrix protein mediated shutdown of host cell metabolism limits vesicular stomatitis virus-induced interferon-alpha responses to plasmacytoid dendritic cells. Immunobiology 212:887–894

Wang T, Town T, Alexopoulou L, Anderson JF, Fikrig E, Flavell RA (2004) Toll-like receptor 3 mediates West Nile virus entry into the brain causing lethal encephalitis. Nat Med 10:1366–1373

Yakub I, Lillibridge KM, Moran A, Gonzalez OY, Belmont J, Gibbs RA, Tweardy DJ (2005) Single nucleotide polymorphisms in genes for 2′-5′-oligoadenylate synthetase and RNase L inpatients hospitalized with West Nile virus infection. J Infect Dis 192:1741–1748

Zhang SY, Jouanguy E, Ugolini S, Smahi A, Elain G, Romero P, Segal D, Sancho-Shimizu V, Lorenzo L, Puel A, Picard C, Chapgier A, Plancoulaine S, Titeux M, Cognet C, von BH, Ku CL, Casrouge A, Zhang XX, Barreiro L, Leonard J, Hamilton C, Lebon P, Heron B, Vallee L, Quintana-Murci L, Hovnanian A, Rozenberg F, Vivier E, Geissmann F, Tardieu M, Abel L,

Casanova JL (2007) TLR3 deficiency in patients with herpes simplex encephalitis. Science 317:1522–1527

Zucchini N, Bessou G, Traub S, Robbins SH, Uematsu S, Akira S, Alexopoulou L, Dalod M (2008) Cutting edge: overlapping functions of TLR7 and TLR9 for innate defense against a herpesvirus infection. J Immunol 180:5799–5803

Zuiderwijk-Sick EA, van der PC, Bsibsi M, Deuzing IP, de BW, Persoon-Deen C, Kondova I, Boven LA, van Noort JM, 't Hart BA, Amor S, Bajramovic JJ (2007) Differentiation of primary adult microglia alters their response to TLR8-mediated activation but not their capacity as APC. Glia 55:1589–1600

Toll-Like Receptors in CNS Parasitic Infections

Bibhuti B. Mishra, Uma Mahesh Gundra, and Judy M. Teale

Contents

Abstract Parasite infections in the central nervous system (CNS) are a major cause of morbidity and mortality worldwide, second only to HIV infection. Finding appropriate therapeutic measures to control CNS parasite infections requires an understanding of the tissue-specific host response. CNS parasitic diseases are invariably associated with persistent T-helper 1 (Th1) cytokine-dependent proinflammatory responses. Although type 1 cytokine-dependent proinflammatory responses are essential to control several types of parasite infections, their persistent production contributes to the development of neuropathology with severe consequences. A family of proteins called Toll-like receptors (TLRs) plays a pivotal role in the induction of inflammatory cytokines during infections and tissue injury. Accumulating evidence indicates that in several CNS parasitic infections such as toxoplasmosis and sleeping sickness, host responses mediated through TLRs contribute to parasite clearance and host survival. However, TLR-mediated responses can also contribute to disease severity, as exemplified in cerebral malaria, neurocysticercosis and river blindness. Thus, TLRs influence the immunopathogenesis of CNS parasitic infections by mechanisms that can either benefit the host or further contribute to CNS pathology. This chapter discusses the immunopathogenesis of parasitic infections in the CNS and the role of TLRs in this process.

B.B. Mishra, U.M. Gundra, and J.M. Teale (✉)
Department of Biology, South Texas Center for Emerging Infectious Diseases,
The University of Texas at San Antonio, One UTSA Circle, San Antonio, TX, 78249-1644, USA
e-mail: Judy.Teale@utsa.edu

T. Kielian (ed.), *Toll-like Receptors: Roles in Infection and Neuropathology*,
Current Topics in Microbiology and Immunology 336,
DOI 10.1007/978-3-642-00549-7_5,

Abbreviations

APC	Antigen-presenting cell
CNS	Central nervous system
CSF	Cerebrospinal fluid
DC	Dendritic cell
GPI	Glycosylphosphatidylinositol
ICAM-1	Intercellular adhesion molecule 1
LNFPIII	Lacto-*N*-fucopentose III
Lyso-PS	Lysophosphatidylserine
MyD88	Myeloid differentiation factor 88
MAL/TIRAP	MyD88-adaptor-like Toll interleukin-1-associated protein
NCC	Neurocysticercosis
NO	Nitric oxide
ODN	Oligodeoxynucleotide
PAMP	Pathogen-associated molecular pattern
PC	Phosphorylcholine
PfEMP-1	*Plasmodium falciparum* erythrocyte membrane protein 1
PFTG	Profilin-like proteins of *Toxoplasma gondii*
PRR	Pattern recognition receptor
RNI	Reactive nitrogen intermediates
ROI	Reactive oxygen intermediates
ROS	Reactive oxygen species
Th1	T-helper 1
Th2	T-helper 2
Th3	T-helper 3
TIR	Toll interleukin-1 receptor
TLR	Toll-like receptor
TRIF	Toll-receptor-associated activator of interferons
TRAM	Toll-receptor-associated molecule
VSG	Variant-specific surface glycoproteins

1 CNS Parasitic Infections and Burden on Society

Although many parasitic infections remain endemic in human populations regardless of geography, they are especially prevalent in developing countries in the tropics. Effective treatments are available for only a few parasitic infections. Sadly, most of these diseases can be avoided by proper sanitation, safe food and drinking water. Thus, parasitic infections have a devastating impact on global health and economic development. With the ever-increasing migration of people and goods between economically developed and developing countries, many parasitic infections have also emerged in developed countries as well.

Mammals have adapted to and co-evolved with parasites for millions of years. Humans harbor over 300 parasite species, although only a few of them cause disease (Cox 2002). Human parasites are classified into six major divisions: Protozoa (amoebae, flagellates, ciliates, sporozoans, coccidia, microsporidia); Nematoda or roundworms; Platyhelminthes or flatworms (cestodes, trematodes); pentastomids or tongue worms; Acanthocephala or thorny-headed worms; and Arthropoda (e.g., insects, spiders, mites, ticks). Among these, protozoans and helminths account for all of the life-threatening parasitic infections in humans.

Across species, parasites generally have distinct but complex life cycles and tissue tropisms within their hosts. Parasitic infections typically produce long-lasting chronic infections in the host. The severity of the infection depends on the infecting dose and/or the number of parasites acquired over a period of time, and most importantly the organs and tissue types affected. Some macroscopic parasites can cause pathology by mechanical disruption as they migrate through or displace tissue. In addition, large, macroscopic parasites can physically obstruct the flow/migration of host substances e.g., cerebral spinal fluid (CSF), thereby enhancing disease severity. Thus, the physical location of the parasite in the host is an important determinant of the infection. Importantly, systemic infections by several particular parasites are not lethal to the host, However infection of the CNS often proves to be fatal. For example, four species of malaria parasites (*Plasmodium falciparum*, *P. malariae*, *P. ovale*, *P. vivax*) infect 300–500 million people every year. Of these, *P. falciparum* causes only 1–2% of all infections but leads to cerebral malaria and more than 95% of the related fatalities. Even survivors often display devastating neurological effects. Other parasitic infections that have similar destructive consequences in the CNS include African trypanosomiasis, neurocysticercosis and amoebic encephalitis. A detailed synopsis of CNS parasitic diseases, annual incidences of infection, and mortality rates are presented in Table 1.

2 Immunopathogenesis of Parasitic Infections

Parasites can be classified into two distinct subsets, namely intracellular parasites and extracellular parasites. Accordingly, hosts employ distinct responses to control both classes of organisms. Nevertheless, protective immunity is a complex process involving both the innate and adaptive immune systems.

In the case of intracellular parasites, like many of the protozoans, effective host resistance depends upon the host cell type. For example, interactions of pathogens with macrophages leads to an oxidative burst and nitric oxide (NO) release that plays an important role in killing the organism. However, when the pathogens encounter other cell types such as dendritic cells (DCs) and fibroblasts that are not good producers of reactive oxygen/nitrogen species, cell-mediated immunity involving MHC Class I and $CD8^+$ T cells is critical for protection. Induction of proinflammatory cytokines such as TNF-α, IL-1, and IL-12 is important for the generation of reactive nitrogen species (RNI) such as NO, either directly or indirectly, through the formation of a Th1 cytokine environment including IFN-γ. Parasites have evolved and developed strategies to utilize host cell machinery to

Table 1 Worldwide impact of CNS parasitic infections

	Disease	Organism	Localization	Incidence	Mortality
Protozoa	Encephalitis (2000)	*T. gondii*	Intracellular	Half a billion worldwide	10 million
	Cerebral malaria (2005)	*P. falciparum*		350–500 million	1.5–2.7 million
	Brain abscess (1998)	*Entamoeba histolytica*	Extracellular	50% of population in tropics	<100,000
	Sleeping sickness (2005)	*Trypanosoma rhodesiense*	Extracellular	50,000–70,000/year	<30,000
		Trypanosoma gambiense		40,000/year	<20,000
	Meningoencephalitis blindness, amoebic encephalitis (2004)	*Acanthamoeba* spp.	Extracellular	100,000/year	<5,000
		Naegleria fowleri		Total of 200 up to 2004	90%
		Balamuthia mandrillaris	Intracellular	Total of 1,500 cases worldwide	95%
Helminth (cestodes)	Neurocysticercosis (2002)	*Taenia solium*	Extracellular	2.5 million worldwide	
	Cerebral hydatid (2000)	*Echinococcus granulosus*		5% worldwide	~200/100,000 infected
Helminth (nematodes)	Meningitis, abscesses (1989)	*Strongyloides stercoralis*	Extracellular	30 million people in 70 countries	~87%
	Epilepsy, blindness (1995)	*Toxocara* spp.		20% of population worldwide	~10%
		Ascaris lumbricoides		1.4 billion total infected	80,000–100,000
	Worm crosses eye, blindness (1997)	*Loa loa*	Intracellular	13 million people worldwide	<15,000
	Eosinophilic meningitis (2003)	*Gnathostoma* sp.		15,000–20,000	~3,500
	Meningoencephalitis (2003)	*Angiostrongylus* spp.	Extracellular	2,500–5,000	<500
		Baylisascaris sp.		~100 cases	10%
	Brain cysts (2003)	*Paragonimus* spp.		20 million worldwide	~5%
Helminth (trematodes)	Seizures, paralysis, spinal cord inflammation (2000)	*Fasciola* spp.	Extracellular	2.4 million worldwide	~1%
		Schistosoma spp.		<200 million people worldwide	15,000 per year

Years given in parentheses represent the date up to which the data were taken or updated.

* Epidemiological data taken from WHO (see http://www.who.int/topics/epidemiology/en/ and http://www.who.int/healthinfo/global_burden_disease/en/index.html).

evade detection and/or avoid harmful mediators of the well-orchestrated host immune response. Some of these strategies include, but are not limited to, entering host cells through alternate pathways, persisting within immune-incompetent cell types, and/or an affinity for immune-privileged tissues. For example, after entering the cell, *Toxoplasma gondii* sequesters itself into a vacuole thereby avoiding lysosomal fusion and acidification. In the case of malarial parasite infections, the disease develops when the parasites manage to enter erythrocytes which lack receptors for MHC Class I and II and IFN-γ, resulting in their inability to present antigens to T-cytotoxic or T-helper cells and to sense available IFN-γ (initial response to malaria is of the Th1 type) to generate a RNI response, respectively. The effector molecules induced by Th1 cytokine responses and cytolytic factors such as granzyme and perforin produced by cytotoxic T cells are nonspecific in nature and can damage the host cells as well. Thus, the persistence of inflammatory processes in intracellular parasite infections often results in tissue damage. This outcome of tissue damage in the CNS results in lethal consequences.

Extracellular parasites vary considerably in size, ranging from microscopic to very large macroscopic species. Host resistance to extracellular parasites involves directly killing them by producing toxic substances, containing the migration of the parasite, and even at times isolating them in particular locations by building barriers. In contrast to intracellular parasites, the preferential development of a Th2 response appears to be an important mechanism of host resistance against extracellular parasites. The production of Th2 type anti-inflammatory cytokines, such as IL-4, 5, 10 and 13, stimulates the formation of IgE and IgG1-type immunoglobulins that are crucial for antiparasitic activity. In addition, these cytokines play an important role in tissue restoration mechanisms that are essential for recovering from mechanical disruption of tissue caused by the size and/or metabolic activity of helminth organisms. Their larger size and/or extracellular locations make it necessary for these parasites to employ unique mechanisms to evade the host immune response. Common evasive mechanisms include varying antigenic surface proteins and masking or releasing tegument target molecules. In addition, helminths actively secrete carbohydrate-containing molecules that lead to immunological tolerance/suppression. However, in such cases, the death of the parasite leads to the loss of active immune suppression and the onset of overwhelming inflammatory responses resulting in severe neurological consequences and death. Collectively, these facts emphasize that the early immune responses during parasitic infections are important for host survival.

3 TLRs in CNS Parasite Infection

The last decade of research has demonstrated that Toll-like receptors of the innate immune system play an important role in the recognition of distinct, mutation-resistant pathogen-associated molecular patterns (PAMPs) of bacteria, viruses, fungi and parasites (Takeda et al. 2003; Heil et al. 2004; Zhang et al. 2004; Gorden et al. 2005; Yarovinsky et al. 2005). Ligand recognition by TLRs invariably culminates in the manifestation of the inflammatory response and induction of adaptive immune responses (Iwasaki and Medzhitov 2004; Takeda and Akira 2005). In addition to PAMPs, TLRs can also be activated by endogenous compounds in various pathological conditions (Leadbetter et al. 2002; Heil et al. 2004; Tsan 2006; Apetoh et al. 2007).

To date, 13 mammalian TLR paralogs have been identified (10 in humans and 13 in mice). The TLR family of receptors are type-I transmembrane proteins with extracellular leucine-rich repeat domains and a cytoplasmic domain with significant homology to the IL-1 receptor type-I domain called the Toll/IL-1 receptor (TIR) (Gay and Keith 1991; Bowie et al. 2004). Once engaged, signaling through TLRs initiates from the TIR domain (Medzhitov 2001) and involves one of the four known adaptor proteins: MyD88 (myeloid differentiation factor 88), MAL/TIRAP (MyD88-adaptor-like/TIR-associated proteins), TRIF (Toll-receptor-associated activator of interferons) and TRAM (Toll-receptor-associated molecule). With the exception of TLR3, all TLRs signal through a common pathway, the MyD88-dependent pathway (Rock et al. 1998). TLR ligation leads to subsequent downstream activation of the NF-κB and MAPK pathways and induces a Th1 proinflammatory response (McDermott and O'Neill 2002). Until very recently, it was thought that TLRs were only essential for the development of Th1 responses. Emerging evidence now suggests that they might be involved in the development of Th2-associated responses as well (Didierlaurent et al. 2004; Dillon et al. 2004; Redecke et al. 2004; Piggott et al. 2005). Nevertheless, the TLR family of molecules play a profound role in eliciting host inflammatory responses.

As discussed earlier, during CNS parasite infection and even when the infection is cured, a persistent inflammatory response may cause irreparable damage and permanent disruption of normal body functions. For example, it is the death of the helminth *Taenia solium*, the causative agent of neurocysticercosis, that generates severe pathology when the active suppression of the immune response by the live parasite is lost. Studies suggesting the important role of TLRs in the immunopathogenesis of CNS parasitic diseases are beginning to emerge. In the first part of this section we discuss the role of TLRs in protozoa-mediated CNS diseases including cerebral malaria and toxoplasmosis which are caused by intracellular protozoa, and sleeping sickness caused by extracellular protozoa. In the second part we discuss the role of TLRs in neurocysticercosis and river blindness, two extracellular helminth parasite-mediated diseases.

3.1 Role of TLRs in CNS Protozoa Infections

3.1.1 Cerebral Malaria

Malaria is caused by four species of protozoan parasites (*P. falciparum, P. malariae, P. ovale, P. vivax*). The malarial life cycle is complex, with multiple stages within the mosquito (vector), as well as humans (primary host), consisting of both extracellular and intracellular phases. Cerebral malaria is caused by the *P. falciparum* species and is characterized by a distinct phenotype of accumulated parasitized red blood cells (RBCs) in cerebral capillaries and venules. The disease is associated with a strong Th1 type inflammatory response consisting of TNF-α, IL-1, IFN-γ, ROI and NO (Gazzinelli and Denkers 2006). The Th1 response is essential to control the early stage of infection when parasites are present in macrophages and hepatocytes; however, the proinflammatory response leads to upregulation of endothelial cell adhesion molecules, including ICAM-1 in cerebral blood vessels (Bauer et al. 2002; van der Heyde et al. 2006). The interaction of ICAM-1 with the *P. falciparum* erythrocyte membrane protein 1 (PfEMP-1), a parasite-encoded factor transported and exposed on the surface of the parasitized erythrocytes, leads to sequestration of parasites in the brain vascular bed causing cerebral malaria (Gitau and Newton 2005). Thus, the Th1 cytokine response elicited during the later stage of *P. falciparum* infection is a primary cause for disease development. This is supported by the fact that IL-4–/– and IL-10–/– mice develop severe malaria along with increased IFN-γ expression (Sanni et al. 2004). It is clear that members of the TLR family are primary inducers of the Th1 inflammatory cytokine response. Indirect evidence for the role of TLRs during human malaria comes from studies where genetic variations for TLR signaling pathways were analyzed and related to disease susceptibility. A polymorphism in TLR4 and MAL/TIRAP was found to influence susceptibility to cerebral malaria (Mockenhaupt et al. 2006). Natural infections in humans are associated with higher expression levels of TLRs 1, 2, 4, and 8, and reduced levels of TLRs 3 and 5 (Ockenhouse et al. 2006; Loharungsikul et al. 2008). However, direct evidence for the involvement of TLRs in malaria comes from studies performed using a murine model of cerebral malaria. MyD88-deficient mice exhibited a reduced parasite load in brain blood vessels and survived the infection as compared to TRIF-deficient and wild-type animals (Coban et al. 2007). Further analysis showed that TLRs 2 and 9, but not TLRs 4, 5, and 7, were responsible for the MyD88-dependent response that contributed to disease severity. However, in a similar study, infected mice deficient for single TLRs 1, 2, 3, 4, 6, 7, or 9 and the adaptor proteins MyD88, TIRAP, and TRIF were reported to be as sensitive to fatal cerebral malaria as wild-type mice (Togbe et al. 2007; Lepenies et al. 2008). The reason(s) for such discrepancies between these studies concerning the role of MyD88 in cerebral malaria pathogenesis is unknown. TLR-mediated development of lethal cerebral malaria in infected mice, however, was dependent on genetic background. For example, deletion of MyD88 on the susceptible C57BL/6 background (a Th1-biased strain) resulted in resistance to cerebral malaria, whereas deletion of MyD88 on the resistant BALB/c background (a Th2-biased strain) increased susceptibility to infection (Griffith et al. 2007). The data suggest that MyD88 modulates the production of immune regulatory cytokines, including inflammatory IFN-γ, IL-17 and anti-inflammatory IL-4, as well as the total number of Foxp3 + regulatory T cells (T_{reg}). On the susceptible Th1-permissive C57BL/6 background, enhanced survival after MyD88 deletion was correlated with a reduction in Th1 cytokines such as IFN-γ and TNF-α. In contrast, on the resistant Th2-permissive BALB/c background, MyD88 deletion led to a very large increase in the production of the Th2 cytokine IL-4, with limited effect on IFN-γ production. However, the role of such a robust IL-4 response to the pathogenesis of cerebral malaria in MyD88-deficient BALB/c mice remains unclear. Another possibility is that increased susceptibility of naturally resistant Th2-permissive BALB/c mice in the absence of MyD88 signaling could be due to its effect on Foxp3⁺ T_{reg} cells. The protected BALB/c mice have a significantly higher expansion of Foxp3⁺ T_{reg} cells than do susceptible C57BL/6 mice during infection. In contrast, the total proportion of Foxp3⁺ T_{reg} cells to the total T cell number decreases in WT C57BL/6 mice compared with that in BALB/c mice. However, MyD88–/– C57BL/6 mice have a significantly increased expansion of Foxp3⁺ T_{reg} cell numbers compared with WT C57BL/6 mice. One possible mechanism to account for these findings would involve TLR-induced Th1 responses that downregulate the expansion of T_{reg} cells, thus contributing to the susceptibility of C57BL/6 WT mice to cerebral malaria. The implication of this explanation is that TLR-mediated Th1 cytokine-dependent responses produced early help to contain the infection, whereas during later stages led to the development of severe cerebral malaria. Support for the involvement of TLRs in the development of cerebral malaria is further strengthened by the fact that the glycosylphosphatidylinositol (GPI) moiety of *P. falciparum*, an important PAMP for TLR2/TLR1 and TLR4 (Krishnegowda et al. 2005; Ropert et al. 2008), alone is sufficient to cause symptoms similar to acute malaria infection (Schofield and Hackett 1993). GPI appears to be necessary and sufficient for the induction of Th1 inflammatory cytokine responses to the parasites and upregulation of ICAM-1 on the surface of cerebral endothelial cells (Schofield and Hackett 1993; Schofield et al. 1996, 2002). Furthermore, antibodies to GPI block the induction of proinflammatory gene expression in target cells of *P. falciparum* (Schofield and Hackett 1993; Schofield et al. 1996, 2002). More importantly, immunization with GPI reverses susceptibility to cerebral malaria (Schofield et al. 2002).

Recently, it has been suggested that TLR9 is also involved in the recognition of parasite factors (Pichyangkul et al. 2004; Coban et al. 2005; Parroche et al. 2007). Hemozoin, a metabolite of hemoglobin produced by *P. falciparum*, activates DCs to produce inflammatory cytokines in a TLR9-dependent manner (Coban et al. 2005). In addition to hemozoin, parasite DNA appears to activate TLR9 to produce proinflammatory cytokines in DCs (Parroche et al. 2007). Taken together, these data indicate a strong role for TLRs in the progression of cerebral malaria.

3.1.2 Toxoplasmosis

During *Toxoplasma gondii* infection, after oral ingestion, the oocyst actively invades the epithelial cell lining of the intestine and disseminates throughout the host, colonizing distant tissues including the brain. The infection is often maintained in a latent stage in the form of intracellular tissue cysts in immune-privileged organs like the CNS. A strong Th1 immune response consisting of IFN-γ and TNF-α expression and the subsequent release of reactive oxygen and nitrogen intermediates (RNI), in particular NO, is essential for protective immunity during

Toxoplasma gondii infection (Bohne et al. 1994; Scharton-Kersten et al. 1997; Yap and Sher 1999; Cai et al. 2000; Halonen and Weiss 2000).

The generation of a Th1 effector response after *Toxoplasma gondii* infection is strictly MyD88 dependent (Scanga et al. 2002; Yarovinsky et al. 2005). Accordingly, in a murine model, MyD88–/– mice were highly susceptible to toxoplasma infection and died very early, whereas wild-type mice survived the infection (Yano et al. 2002). In MyD88–/– mice, the simultaneous absence of IFN-γ and NO is marked by uncontrolled parasite replication and widespread dissemination with higher loads in brain tissue. In addition to MyD88-dependent IL-12 production by phagocytes (Jankovic et al. 2002; Scanga et al. 2002), involvement of MyD88 signaling in T cells is necessary for immunity to toxoplasma. Chimeric mice lacking MyD88 in T cells (*MyD88–/–*) display uncontrolled parasite growth associated with lesions in the brain parenchyma and early death (LaRosa et al. 2008). However, it is unknown whether or not TLRs are directly involved in the MyD88-dependent response by phagocytes and/or T cells.

TLRs 1, 2, 4, 6, 9 and 11, all of which activate MyD88 signaling individually, are not sufficient to control *Toxoplasma gondii* infection (Scanga et al. 2002; Vossenkamper et al. 2004; Hitziger et al. 2005; Minns et al. 2006; Debierre-Grockiego et al. 2007), as mice deficient in either of them display similar survival rates compared to wild-type mice when infected with toxoplasma (Debierre-Grockiego et al. 2007). Nonetheless, TLR11–/– and TLR2/4 double knockout mice display relatively increased susceptibility to infection with a simultaneous decrease in IL-12 along with an increase in the number of brain cysts (Yarovinsky et al. 2005; Debierre-Grockiego et al. 2007). In addition, TLR2–/– mice die early as compared to the wild type when infected with higher doses of *Toxoplasma gondii* (Mun et al. 2003). Thus, it is not fully clear which TLR(s) alone or in combination with other types of pattern recognition receptors (PRRs) are involved in the recognition of toxoplasma-derived molecules and provide the MyD88-dependent mechanism for the resistance to this infection. Unlike either MyD88–/– or IL-12–/– mice (Scanga et al. 2002), TLR-deficient animals retain a residual amount of IFN-γ production that is most likely sufficient to provide protection against the acute phase of infection (Yarovinsky et al. 2005). Nevertheless, the identification of several toxoplasma-derived molecules that trigger a TLR-dependent upregulation of TNF-α and IL-12 synthesis (Yarovinsky and Sher 2006) strongly suggest a role for TLRs in host immunity to this infection.

Glycosylphosphatidylinositols (GPI) lipid anchors and profilin-like proteins are highly conserved molecules present in a number of protozoans, including *Toxoplasma gondii*. GPI-anchored proteins are the dominant factors expressed on the surface of *Toxoplasma gondii* (Black and Boothroyd 2000; Lekutis et al. 2001). GPIs isolated from *Toxoplasma gondii*, as well as the core glycan and lipid moieties, activate both TLR4 and TLR2 for cytokine production (Debierre-Grockiego et al. 2007). Recently, profilin-like proteins of *Toxoplasma gondii* (PFTG) were found to be essential for the parasite's gliding motility, host cell invasion, active egress from host cells, and virulence in mice (Plattner et al. 2008). Interestingly, the induction of IL-12 in DCs exposed to PFTG was abolished in

TLR11-deficient mice. Taken together, increased susceptibility and the number of brain cysts during *Toxoplasma gondii* infection of TLR11–/– mice strongly suggests a role for TLR11 in the protection mechanism (Yarovinsky et al. 2005). However, the TLR11 gene in humans has a premature stop codon and is not functional (Zhang et al. 2004), thereby suggesting the involvement of additional PAMPs and PRRs in the protection mechanism. In this regard, it is worth noting that other *Toxoplasma gondii* molecules, such as tachyzoite heat-shock proteins and other partially purified tachyzoite preparations, activate TLR4 and TLR2, respectively (Aosai et al. 2002; Del Rio et al. 2004). These data indicate a strong role for TLRs in the protection mechanism involved in CNS toxoplasmosis.

3.1.3 Sleeping Sickness

Sleeping sickness is a parasitic disease caused by the extracellular protozoan African trypanosome duo *Trypanosoma rhodesiense* and *Trypanasoma gambiense* (see Table 1). Trypanosomes circulate in the bloodstream, lymphatics, and cerebrospinal fluid (CSF) but are not thought to cross the BBB and enter the CNS parenchyma (Sternberg 2004). However, the parasites have been detected in some CNS areas where the BBB is not fully developed, such as the circumventricular organs, dorsal root ganglia and hypothalamus. Nevertheless, parasites are present in cerebral blood vessels even early during infection. Parasite clearance during the early stages of infection depends on the development of a Th1 cytokine environment, specifically IFN-γ activity against the parasite variant-specific surface glycoprotein (VSG) (Hertz et al. 1998). The development of an antibody response against VSG is required to clear the parasite in the later stages of infection and/or during secondary infections (Reinitz and Mansfield 1988; Theodos et al. 1990; Schleifer et al. 1993).

Recent evidence suggests that manifestation of the Th1 cytokine response during *Trypanosoma* infection is MyD88 dependent (Drennan et al. 2005). After *Trypanosoma brucei* infection, MyD88-deficient mice express reduced IFN-γ and TNF levels concurrent with elevated parasitemia in the blood (Drennan et al. 2005). These mice are more susceptible to infection when compared with their wild-type counterparts. MyD88 signaling also appears to influence humoral immunity, as infection in MyD88-deficient mice was associated with lower VSG-specific IgG2a antibody titers. This MyD88-dependent response in infected animals is partially dependent on TLR9.

Treatment with CpG oligodeoxynucleotide (ODN), a TLR9 agonist, promotes protection against African trypanosomes in otherwise susceptible animals (Harris et al. 2007). VSG-specific antibodies increased in the CpG-ODN-treated infected mice; however, the TLR9-specific ligand in African trypnosoma is unknown. DNA from a related organism, American trypanosome (*Trypanosoma. cruzi*), stimulates Th1 cytokine production by antigen-presenting cells (APCs) in a TLR9-dependent manner (Bafica et al. 2006). By extension, it is likely that African trypnosome genomic DNA might be a TLR9 ligand. The DNA of *Trypanosoma brucei* and

Trypanosoma rhodesiense activates macrophages to produce NO, prostaglandin E_2, and the cytokines IL-6, IL-10, IL-12, and TNF-α, and it activates a cascade of signaling molecules involved in downstream signaling of TLRs (Harris et al. 2006). Besides TLR9, other TLRs that might participate in the MyD88-dependent generation of a protective Th1 response during African trypanosome infection are not known.

In protozoa infections, an important set of PAMPs are the GPI-anchored surface molecules, which induce a very high inflammatory response through TLR2 and TLR4 in conjunction with MyD88 (Gazzinelli and Denkers 2006). However, in the case of African trypanosome, the involvement of TLRs in the host response to GPI factors is not yet clear. The variant-specific surface glycoproteins (VSGs) are the major GPI-anchored structures present on African trypanosomes. The GPI residues of the membrane-bound VSGs (mVSGs) induce macrophages to produce TNF-α and IL-6 in a MyD88-dependent manner (Drennan et al. 2005), while the production of IL-12p40 and NO is MyD88 independent. In contrast, macrophage production of inflammatory cytokines in response to soluble VSG (sVSG)-induced activation of the NF-κB cascade is independent of a TLR-mediated interaction (Leppert et al. 2007).

Evidence shows that signaling through MyD88, which is a primary adaptor molecule for most TLRs, in addition to the IL-1R and IL-18R, plays an important role during protozoa parasite infection in CNS. As discussed earlier, it seems that disease containment (toxoplasmosis and sleeping sickness) or exacerbation (cerebral malaria) seems to depend upon the same Th1 proinflammatory response induced through the use of MyD88 signaling during the course of a particular disease. While our knowledge about the role of TLRs in resistance or enhanced pathology to protozoan parasite infection in CNS is limited, there is growing awareness about their involvement in CNS immunopathogenesis during these infections. Activation of TLRs by the abovementioned protozoan molecules leads to the expression of proinflammatory cytokines (for example IL-12) that promote Th1 responses. Thus, exploitation of TLR signaling using the protozoan-derived TLR ligands holds promise as possible therapeutics.

3.2 Role of TLRs in CNS Helminth Infections

3.2.1 Neurocysticercosis (NCC)

Neurocysticercosis is the most common parasitic disease of the CNS caused by the larvae of *Taenia solium* (White 1997, 2000). The disease has a long asymptomatic phase in humans characterized by the absence of a detectable inflammatory response surrounding the parasite (White et al. 1997). The overall immune response during the asymptomatic phase is of the Th2 phenotype (Chavarria et al. 2003). However, the destruction of larvae by therapeutic treatment or by normal parasite attrition causes a strong inflammatory response, often consisting of a chronic granulomatous reaction and manifestation of typical symptoms of the disease (White et al. 1997). The immune response in the CNS of symptomatic patients consists of an overt Th1 phenotype (Restrepo et al. 1998) or a mixed Th1, Th2, and Th3 response, depending upon the absence or presence of granulomas (Restrepo et al. 2001). The hyperinflammatory response prevailing during the symptomatic phase in the CNS is responsible for the severe neuropathology and mortality associated with neurocysticercosis (White 1997). In a murine model of neurocysticercosis, mice infected intracranially (i.c.) with *Mesocestoides corti* elicit a predominant Th1 cytokine response (IL-12 and IFN-γ) throughout the duration of infection (13 weeks) (Cardona et al. 1999). The Th1 cytokine response is mounted by γδ T cells (Cardona and Teale 2002) that contribute to infiltration of immune cells into the brain (Cardona et al. 2003) and enhance neuropathology and disease severity (Cardona and Teale 2002). Thus, in the chronic stage of both human and murine neurocysticercosis, manifestation of the Th1 inflammatory response is the main cause of disease severity and mortality.

In murine NCC, all known TLRs (TLRs 1–13) except TLR5 are upregulated and differentially expressed among various nervous tissue and infiltrating immune system cell types (see Table 2) (Mishra et al. 2006, 2008). Among the infiltrating cells both macrophages and γδ T cells express the majority of the TLRs. Our unpublished data suggest that MyD88-deficient mice infected with *M. corti* display an enhanced ability to survive the infection as compared to wild-type mice, although the numbers of parasites remain the same. These mice exhibit decreased Th1 inflammatory cytokine responses along with reduced neuropathology. The increased resistance of MyD88-deficient mice coincides with a reduction in the infiltration of immune cells, including γδ T cells, particularly during early stages of infection (< 3 weeks p.i.). γδ T cells from NCC mice produce a high level of proinflammatory cytokines and chemokines in response to TLR ligands, suggesting a role for TLRs in the severity of neurocysticercosis (unpublished data). However, it has not yet been determined whether TLRs are directly involved in the MyD88-dependent responses that contribute to pathology in NCC, since this adaptor also transduces signals via the IL-1R and IL-18R.

Among all the TLRs, TLR2 expression in both the uninfected and NCC brain is highest in terms of the number of TLR2-positive nervous tissue cells as well as level of expression (Mishra et al. 2006). In addition, TLR2 is the first to be upregulated after parasite infection, particularly in astrocytes and their foot processes that terminate on blood vessels present in the periventricular and meningeal areas (Mishra et al. 2006). During NCC, astrocytic expression of IL-6 in proximity to pial vessels may compromise the blood–brain barrier (Alvarez and Teale 2006); these vessels are the primary infiltration site of immune cells into the brain (Alvarez and Teale 2006, 2007b,a). Immunopathological studies indicated that TLR2-deficient mice display reduced inflammatory responses, increased parasite burden and increased neuropathology. Thus, the TLR2-dependent response contributes to host protection during NCC. It is possible that TLR2 is selectively involved in immune regulatory mechanisms that might help to control neuroinflammation during NCC (Netea et al. 2004; Nylen and Sacks 2007). Nevertheless, the helminth-derived

Table 2 TLRs in pathogenesis of CNS parasitic infections

Disease		Disease severity	Phenotype	Ligands; TLRs
Cerebral malaria	MyD88	Decreased susceptibility (Coban et al. 2007; Griffith et al. 2007)	Reduced TH1 response and decreased parasite load in cerebral blood vessels	GPI; TLR2/1 and TLR4 (Krishnegowda et al. 2005; Ropert et al. 2008)
		As susceptible as the wild type (Lepenies et al. 2008; Togbe et al. 2007)	Similar to wild type	Hemozoin; TLR9 (Coban et al. 2005)
	TLRs 2, 9	Decreased susceptibility (Coban et al. 2005)	Partially reduced IL-12 (TLR2–/–) and TNF-α (TLR9–/–)	DNA; TLR9 (Parroche et al. 2007)
		No effect (Lepenies et al. 2008; Togbe et al. 2007)	Similar to wild type	
	TLRs 4, 5, 7	No effect (Coban et al. 2005)	Similar to wild type	
		No effect (Lepenies et al. 2008; Togbe et al. 2007)		
	TLRs 1, 3, 6	No effect (Lepenies et al. 2008; Togbe et al. 2007)	Similar to wild type	
Toxoplasmosis	MyD88	Highly susceptible (LaRosa et al. 2008; Yano et al. 2002)	Reduced IFN-γ and NO and increased parasite burden in CNS	GPI; TLR2 (Debierre-Grockiego et al. 2007; Yarovinsky and Sher 2006)
	TLRs 1, 2, 4	No effect (Debierre-Grockiego et al. 2007; Hitziger et al. 2005; Minns et al. 2006; Scanga et al. 2002; Vossenkamper et al. 2004)	Similar to wild type	Profilin-like protein; TLR11 (Yarovinsky and Sher 2006; Yarovinsky et al. 2005)
	TLR2 (high dose)	Increased susceptibility (Mun et al. 2003)		
	TLRs 2/4 (double knockout)	Slight increase in susceptibility (Debierre-Grockiego et al. 2007)	Increased parasite burden in CNS	
	TLR11 (Yarovinsky et al. 2005)	Increased susceptibility	Reduced IFN-γ and NO and increased parasite burden in CNS	Heat shock proteins; TLR2 and TLR4 (Aosai et al. 2002; Del Rio et al. 2004)

(continued)

Table 2 (continued)

Disease		Disease severity	Phenotype	Ligands; TLRs
Sleeping sickness	MyD88 (Drennan et al. 2005)	Increased susceptibility	Reduced Th1 response	GPI-VSP; TLR2 and 4 (Drennan et al. 2005)
	TLR9 (Drennan et al. 2005)	Increased susceptibility	Reduced Th1 response	DNA; TLR9??
Neurocysticercosis	MyD88 (Mishra et.al, unpublished)	Enhanced survival with decreased pathology	Reduced Th1 response	Not known
	TLR2 (Gundra et.al. unpublished)	Increased susceptibility, severe pathology	Reduced Th1 response	
River blindness/ onchocerciasis	MyD88 (Gillette-Ferguson et al. 2007; Hise et al. 2003, 2007)	No disease or associated pathology	Reduced Th1 and increased Th2 response	GPI; TLR2 (Gillette-Ferguson et al. 2007)
	TLR2 (Gillette-Ferguson et al. 2006)	No disease or associated pathology	Reduced Th1 and increased Th2 response	
	TLRs 4, 9 (Gillette-Ferguson et al. 2006)	As susceptible as the wild type	Similar to wild type	

glycoconjugate lacto-*N*-fucopentose III (LNFPIII) from schistosomes and the filarial phosphorylcholine (PC)-containing secreted glycoprotein ES-62 both stimulate DCs to induce a regulatory Th2 response in a TLR4-dependent fashion (Thomas et al. 2003; Goodridge et al. 2004). Similarly, the TLR2 ligand, zymosan, does not induce a Th1 response; rather it engages both TLR2 and the C-type lectin receptor, dectin-1, to activate DCs and prime IL-10-producing T cells. These T cells, known as T regulatory cells (T_{reg}), do not typically secrete Th1 and Th2 cytokines but rather contribute to immunological tolerance (Dillon et al. 2006). In addition, the activation of TLR2 on DCs by lysophosphatidylserine (lyso-PS) is essential for the development of IL-10-producing T_{reg} cells in schistosome infections (van der Kleij et al. 2002). In fact, TLR2–/– mice infected with *Schistosome mansoni* are susceptible to infection due to an elevated immunopathology along with a failure to expand the T_{reg} population (Layland et al. 2007). It is possible that TLR2 plays a similar regulatory role in NCC by controlling CNS inflammation and protecting the host. Collectively, these observations suggest that TLRs influence the disease outcome of neurocysticercosis.

3.2.2 River Blindness/Onchocerciasis

Onchocerciasis is caused by the filarial nematode *Onchocerca volvulus* (Hoerauf and Brattig 2002). The larvae stage of *O. volvulus* develops into mature worms residing in subcutaneous (s.c.) nodules for 10–15 years and producing millions of microfilaria (Mf) in infected individuals. Mf can migrate through the skin and reach the eye. While the Mf are alive they do not induce any detectable inflammatory response. However, the death/degeneration of Mf in the cornea trigger a local host inflammatory response and cellular infiltration leading to corneal edema and opacification, resulting in blindness (Hall and Pearlman 1999). The major culprits are thought to be the infiltrating leukocytes such as neutrophils and eosinophils that have a cytotoxic effect on resident cells in the cornea, hence triggering the associated pathological sequelae. *O. volvulus*, like many other filarial nematodes, carries endosymbiotic *Wolbachia* bacteria (Taylor and Hoerauf 1999). In a mouse model for river blindness in which parasites or a soluble extract were injected directly into the corneal stroma, the extracts of *O. volvulus* containing *Wolbachia* induced substantial corneal inflammation, cellular infiltration and keratitis. The extract that did not contain *Wolbachia* failed to induce such effects (Saint Andre et al. 2002). It is now clear that *Wolbachia*, the endocytic bacteria, has an essential role in the *O. volvulus*-induced inflammatory response and keratitis–river blindness (Hise et al. 2003).

In the murine model of onchocerciasis/river blindness, neutrophil recruitment to the cornea and development of corneal haze is exclusively MyD88 dependent (Gillette-Ferguson et al. 2006). MyD88–/– mice exhibit an absence of neutrophil chemokines such as CXCL1/KC and CXCL2/MIP-2 in the cornea. Neutrophils from MyD88-deficient mice are completely unresponsive to *Wolbachia* and *O. volvulus* extracts and do not produce CXCL1, CXCL2, or TNF-α. Interestingly, neutrophil recruitment and activation is strictly dependent on TLR2 (Gillette-Ferguson et al. 2007). In addition to neutrophils, resident cells such as corneal fibroblasts as well as infiltrating macrophages and dendritic cells all produce inflammatory cytokines in a TLR2- and MyD88-dependent manner, thereby contributing to the development of the disease (Johnson et al. 2005; Hise et al. 2007). Furthermore, TLR2, in association with MyD88, functions in the development of a systemic Th1 cytokine response (IFN-γ) during the disease, while having no such role in the manifestation of a host Th2-type response (Daehnel et al. 2007). Thus, TLR2 plays a pivotal role in the development of a persistent hyperinflammatory response eventuating in corneal haze, as observed in river blindness.

4 Conclusions and Perspectives

A considerable body of evidence supports the view that TLRs modulate host inflammatory responses that influence the outcome of several deadly CNS parasitic diseases such as cerebral malaria, toxoplasmosis, sleeping sickness, neurocysticercosis and river blindness. Whether the TLR response will lead to disease containment or exacerbation seems to depend upon, but is probably not limited to, the organism class (extracellular vs. intracellular parasites), size (microscopic vs. macroscopic), the duration of the infection and the cell type infected in the case of intracellular parasites. For instance, a MyD88-dependent response is sufficient to control sleeping sickness caused by extracellular protozoa and toxoplasmosis where parasites preferentially invade phagocytes. In the case of malaria, while TLR-mediated Th1-dependent effector molecules help to control the initial erythrocytic phase of infection, the same effector molecules are inefficient in controlling the later stages of infection. In fact, some of the cell adhesion molecules induced, such as ICAM-1, actually facilitate the development of cerebral malaria. Although the TLR(s) responsible for the control of toxoplasmosis are not yet known, some *Toxoplasma gondii* PAMPs are able to elicit a Th1-protective response, implying a role for TLRs acting in concert to control infection. This suggests that TLR-induced responses can effect either disease containment (toxoplasmosis and sleeping sickness) or exacerbation of the disease (cerebral malaria) depending upon the duration of the infection and the tissue tropism of the pathogen.

TLRs are involved in the exacerbation of extracellular helminth parasitic infections in the chronic stage, particularly when the parasite is nonviable. However, we are still far from understanding the complete picture of the individual TLRs involved, the molecules produced by the infectious agents, and the cell types involved. In addition, complex organisms like helminths have evolved together with their symbiotic/commensal association with bacteria that might influence the TLR-mediated response during disease development; a clear example is *Wolbachia* in river blindness. Thus, it will be essential to know the associated microbe's contribution, if any, in the TLR-mediated inflammatory response, particularly in helminth infections.

A critical question that remains is whether TLRs have other functions besides inflammation that may be important in CNS helminth infections. The death of TLR2-deficient mice during experimental NCC, in contrast to the reduced severity in MyD88-deficient mice, suggests a role of TLR2 in the regulatory/anti-inflammatory protective response. Uncontrolled inflammation is detrimental in several CNS parasitic diseases; therefore, determining how TLRs and parasite-derived molecules elicit inflammatory responses versus regulatory/anti-inflammatory responses will be crucial for formulating any therapeutic strategies. Additionally, although a fair amount of knowledge has been acquired regarding the development of the Th1 responses and the essential role of TLRs in this process, how a Th2 response develops and the possible roles of TLRs in this process are not yet clear.

Whether TLRs influence other CNS parasitic infections that cause significant mortality, for example those caused by *Entamoeba histolytica*, *Strongyloides stercoralis* and *Ascaris lumbricoides*, has not yet been studied in detail. Nevertheless, the host probably elicits highly polarized responses depending upon the class of parasite. Therefore, it is reasonable to predict that TLRs might have similar roles in many of these CNS diseases whose phenotypes are similar to the ones discussed in the present review. However, it is essential to test the role of TLRs in the context of immunopathogenesis of these diverse infections so that efficacious therapies can be designed.

Acknowledgments Supported by awards NS 35974, AI 59703 and 1P01AI057986 from the National Institutes of Health.

References

Alvarez JI, Teale JM (2006) Breakdown of the blood brain barrier and blood-cerebrospinal fluid barrier is associated with differential leukocyte migration in distinct compartments of the CNS during the course of murine NCC. J Neuroimmunol 173:45–55

Alvarez JI, Teale JM (2007a) Differential changes in junctional complex proteins suggest the ependymal lining as the main source of leukocyte infiltration into ventricles in murine neurocysticercosis. J Neuroimmunol 187:102–113

Alvarez JI, Teale JM (2007b) Evidence for differential changes of junctional complex proteins in murine neurocysticercosis dependent upon CNS vasculature. Brain Res 1169:98–111

Aosai F, Chen M, Kang HK, Mun HS, Norose K, Piao LX, Kobayashi M, Takeuchi O, Akira S, Yano A (2002) *Toxoplasma gondii*-derived heat shock protein HSP70 functions as a B cell mitogen. Cell Stress Chaperones 7:357–364

Apetoh L, Ghiringhelli F, Tesniere A, Obeid M, Ortiz C, Criollo A, Mignot G, Maiuri MC, Ullrich E, Saulnier P, Yang H, Amigorena S, Ryffel B, Barrat FJ, Saftig P, Levi F, Lidereau R, Nogues C, Mira JP, Chompret A, Joulin V, Clavel-Chapelon F, Bourhis J, Andre F, Delaloge S, Tursz T, Kroemer G, Zitvogel L (2007) Toll-like receptor 4-dependent contribution of the immune system to anticancer chemotherapy and radiotherapy. Nat Med 13:1050–1059

Bafica A, Santiago HC, Goldszmid R, Ropert C, Gazzinelli RT, Sher A (2006) Cutting edge: TLR9 and TLR2 signaling together account for MyD88-dependent control of parasitemia in *Trypanosoma cruzi* infection. J Immunol 177:3515–3519

Bauer PR, Van Der Heyde HC, Sun G, Specian RD, Granger DN (2002) Regulation of endothelial cell adhesion molecule expression in an experimental model of cerebral malaria. Microcirculation 9:463–470

Black MW, Boothroyd JC (2000) Lytic cycle of *Toxoplasma gondii*. Microbiol Mol Biol Rev 64:607–623

Bohne W, Heesemann J, Gross U (1994) Reduced replication of *Toxoplasma gondii* is necessary for induction of bradyzoite-specific antigens: a possible role for nitric oxide in triggering stage conversion. Infect Immun 62:1761–1767

Bowie AG, Zhan J, Marshall WL (2004) Viral appropriation of apoptotic and NF-kappaB signaling pathways. J Cell Biochem 91:1099–1108

Cai G, Radzanowski T, Villegas EN, Kastelein R, Hunter CA (2000) Identification of STAT4-dependent and independent mechanisms of resistance to *Toxoplasma gondii*. J Immunol 165:2619–2627

Cardona AE, Gonzalez PA, Teale JM (2003) CC chemokines mediate leukocyte trafficking into the central nervous system during murine neurocysticercosis: role of gamma delta T cells in amplification of the host immune response. Infect Immun 71:2634–2642

Cardona AE, Restrepo BI, Jaramillo JM, Teale JM (1999) Development of an animal model for neurocysticercosis:immune response in the central nervous system is characterized by a predominance of gamma delta T cells. J Immunol 162:995–1002

Cardona AE, Teale JM (2002) Gamma/delta T cell-deficient mice exhibit reduced disease severity and decreased inflammatory response in the brain in murine neurocysticercosis. J Immunol 169:3163–3171

Chavarria A, Roger B, Fragoso G, Tapia G, Fleury A, Dumas M, Dessein A, Larralde C, Sciutto E (2003) TH2 profile in asymptomatic *Taenia solium* human neurocysticercosis. Microbes Infect 5:1109–1115

Coban C, Ishii KJ, Kawai T, Hemmi H, Sato S, Uematsu S, Yamamoto M, Takeuchi O, Itagaki S, Kumar N, Horii T, Akira S (2005) Toll-like receptor 9 mediates innate immune activation by the malaria pigment hemozoin. J Exp Med 201:19–25

Coban C, Ishii KJ, Uematsu S, Arisue N, Sato S, Yamamoto M, Kawai T, Takeuchi O, Hisaeda H, Horii T, Akira S (2007) Pathological role of Toll-like receptor signaling in cerebral malaria. Int Immunol 19:67–79

Cox FE (2002) History of human parasitology. Clin Microbiol Rev 15:595–612

Daehnel K, Gillette-Ferguson I, Hise AG, Diaconu E, Harling MJ, Heinzel FP, Pearlman E (2007) Filaria/*Wolbachia* activation of dendritic cells and development of Th1-associated responses is dependent on Toll-like receptor 2 in a mouse model of ocular onchocerciasis (river blindness). Parasite Immunol 29:455–465

Debierre-Grockiego F, Campos MA, Azzouz N, Schmidt J, Bieker U, Resende MG, Mansur DS, Weingart R, Schmidt RR, Golenbock DT, Gazzinelli RT, Schwarz RT (2007) Activation of TLR2 and TLR4 by glycosylphosphatidylinositols derived from *Toxoplasma gondii*. J Immunol 179:1129–1137

Del Rio L, Butcher BA, Bennouna S, Hieny S, Sher A, Denkers EY (2004) *Toxoplasma gondii* triggers myeloid differentiation factor 88-dependent IL-12 and chemokine ligand 2 (monocyte chemoattractant protein 1) responses using distinct parasite molecules and host receptors. J Immunol 172:6954–6960

Didierlaurent A, Ferrero I, Otten LA, Dubois B, Reinhardt M, Carlsen H, Blomhoff R, Akira S, Kraehenbuhl JP, Sirard JC (2004) Flagellin promotes myeloid differentiation factor 88-dependent development of Th2-type response. J Immunol 172:6922–6930

Dillon S, Agrawal A, Van Dyke T, Landreth G, McCauley L, Koh A, Maliszewski C, Akira S, Pulendran B (2004) A Toll-like receptor 2 ligand stimulates Th2 responses in vivo, via induction of extracellular signal-regulated kinase mitogen-activated protein kinase and c-Fos in dendritic cells. J Immunol 172:4733–4743

Dillon S, Agrawal S, Banerjee K, Letterio J, Denning TL, Oswald-Richter K, Kasprowicz DJ, Kellar K, Pare J, van Dyke T, Ziegler S, Unutmaz D, Pulendran B (2006) Yeast zymosan, a stimulus for TLR2 and dectin-1, induces regulatory antigen-presenting cells and immunological tolerance. J Clin Invest 116:916–928

Drennan MB, Stijlemans B, Van den Abbeele J, Quesniaux VJ, Barkhuizen M, Brombacher F, De Baetselier P, Ryffel B, Magez S (2005) The induction of a type 1 immune response following a *Trypanosoma brucei* infection is MyD88 dependent. J Immunol 175:2501–2509

Gay NJ, Keith FJ (1991) Drosophila Toll and IL-1 receptor. Nature 351:355–356

Gazzinelli RT, Denkers EY (2006) Protozoan encounters with Toll-like receptor signalling pathways:implications for host parasitism. 6:895–906

Gillette-Ferguson I, Daehnel K, Hise AG, Sun Y, Carlson E, Diaconu E, McGarry HF, Taylor MJ, Pearlman E (2007) Toll-like receptor 2 regulates CXC chemokine production and neutrophil recruitment to the cornea in *Onchocerca volvulus/Wolbachia*-induced keratitis. Infect Immun 75:5908–5915

Gillette-Ferguson I, Hise AG, Sun Y, Diaconu E, McGarry HF, Taylor MJ, Pearlman E (2006) *Wolbachia*- and *Onchocerca volvulus*-induced keratitis (river blindness) is dependent on myeloid differentiation factor 88. Infect Immun 74:2442–2445

Gitau EN, Newton CR (2005) Blood–brain barrier in falciparum malaria (review article). Trop Med Int Health 10:285–292

Goodridge HS, Marshall FA, Wilson EH, Houston KM, Liew FY, Harnett MM, Harnett W (2004) In vivo exposure of murine dendritic cell and macrophage bone marrow progenitors to the phosphorylcholine-containing filarial nematode glycoprotein ES-62 polarizes their differentiation to an anti-inflammatory phenotype. Immunology 113:491–498

Gorden KB, Gorski KS, Gibson SJ, Kedl RM, Kieper WC, Qiu X, Tomai MA, Alkan SS, Vasilakos JP (2005) Synthetic TLR agonists reveal functional differences between human TLR7 and TLR8. J Immunol 174:1259–1268

Griffith JW, O'Connor C, Bernard K, Town T, Goldstein DR, Bucala R (2007) Toll-like receptor modulation of murine cerebral malaria is dependent on the genetic background of the host. J Infect Dis 196:1553–1564

Hall LR, Pearlman E (1999) Pathogenesis of onchocercal keratitis (river blindness). Clin Microbiol Rev 12:445–453

Halonen SK, Weiss LM (2000) Investigation into the mechanism of gamma interferon-mediated inhibition of *Toxoplasma gondii* in murine astrocytes. Infect Immun 68:3426–3430

Harris TH, Cooney NM, Mansfield JM, Paulnock DM (2006) Signal transduction, gene transcription, and cytokine production triggered in macrophages by exposure to trypanosome DNA. Infect Immun 74:4530–4537

Harris TH, Mansfield JM, Paulnock DM (2007) CpG oligodeoxynucleotide treatment enhances innate resistance and acquired immunity to African trypanosomes. Infect Immun 75:2366–2373

Heil F, Hemmi H, Hochrein H, Ampenberger F, Kirschning C, Akira S, Lipford G, Wagner H, Bauer S (2004) Species-specific recognition of single-stranded RNA via toll-like receptor 7 and 8. Science 303:1526–1529

Hertz CJ, Filutowicz H, Mansfield JM (1998) Resistance to the African trypanosomes is IFN-gamma dependent. J Immunol 161:6775–6783

Hise AG, Daehnel K, Gillette-Ferguson I, Cho E, McGarry HF, Taylor MJ, Golenbock DT, Fitzgerald KA, Kazura JW, Pearlman E (2007) Innate immune responses to endosymbiotic *Wolbachia* bacteria in *Brugia malayi* and *Onchocerca volvulus* are dependent on TLR2, TLR6, MyD88, and Mal, but not TLR4, TRIF, or TRAM. J Immunol 178:1068–1076

Hise AG, Gillette-Ferguson I, Pearlman E (2003) Immunopathogenesis of *Onchocerca volvulus* keratitis (river blindness):a novel role for TLR4 and endosymbiotic *Wolbachia* bacteria. J Endotoxin Res 9:390–394

Hitziger N, Dellacasa I, Albiger B, Barragan A (2005) Dissemination of *Toxoplasma gondii* to immunoprivileged organs and role of Toll/interleukin-1 receptor signalling for host resistance assessed by in vivo bioluminescence imaging. Cell Microbiol 7:837–848

Hoerauf A, Brattig N (2002) Resistance and susceptibility in human onchocerciasis: beyond Th1 vs. Th2. Trends Parasitol 18:25–31

Iwasaki A, Medzhitov R (2004) Toll-like receptor control of the adaptive immune responses. Nat Immunol 5:987–995

Jankovic D, Kullberg MC, Hieny S, Caspar P, Collazo CM, Sher A (2002) In the absence of IL-12, CD4(+) T cell responses to intracellular pathogens fail to default to a Th2 pattern and are host protective in an IL-10(–/–) setting. Immunity 16:429–439

Johnson AC, Heinzel FP, Diaconu E, Sun Y, Hise AG, Golenbock D, Lass JH, Pearlman E (2005) Activation of toll-like receptor (TLR)2, TLR4, and TLR9 in the mammalian cornea induces MyD88-dependent corneal inflammation. Invest Ophthalmol Vis Sci 46:589–595

Krishnegowda G, Hajjar AM, Zhu J, Douglass EJ, Uematsu S, Akira S, Woods AS, Gowda DC (2005) Induction of proinflammatory responses in macrophages by the glycosylphosphatidylinositols of *Plasmodium falciparum*: cell signaling receptors, glycosylphosphatidylinositol (GPI) structural requirement, and regulation of GPI activity. J Biol Chem 280:8606–8616

LaRosa DF, Stumhofer JS, Gelman AE, Rahman AH, Taylor DK, Hunter CA, Turka LA (2008) T cell expression of MyD88 is required for resistance to *Toxoplasma gondii*. Proc Natl Acad Sci USA 105:3855–3860

Layland LE, Rad R, Wagner H, da Costa CU (2007) Immunopathology in schistosomiasis is controlled by antigen-specific regulatory T cells primed in the presence of TLR2. Eur J Immunol 37:2174–2184

Leadbetter EA, Rifkin IR, Hohlbaum AM, Beaudette BC, Shlomchik MJ, Marshak-Rothstein A (2002) Chromatin-IgG complexes activate B cells by dual engagement of IgM and Toll-like receptors. Nature 416:603–607

Lekutis C, Ferguson DJ, Grigg ME, Camps M, Boothroyd JC (2001) Surface antigens of *Toxoplasma gondii*: variations on a theme. Int J Parasitol 31:1285–1292

Lepenies B, Cramer JP, Burchard GD, Wagner H, Kirschning CJ, Jacobs T (2008) Induction of experimental cerebral malaria is independent of TLR2/4/9. Med Microbiol Immunol 197:39–44

Leppert BJ, Mansfield JM, Paulnock DM (2007) The soluble variant surface glycoprotein of African trypanosomes is recognized by a macrophage scavenger receptor and induces I kappa B alpha degradation independently of TRAF6-mediated TLR signaling. J Immunol 179: 548–556

Loharungsikul S, Troye-Blomberg M, Amoudruz P, Pichyangkul S, Yongvanitchit K, Looareesuwan S, Mahakunkijcharoen Y, Sarntivijai S, Khusmith S (2008) Expression of toll-like receptors on antigen-presenting cells in patients with falciparum malaria. Acta Trop 105:10–15

McDermott EP, O'Neill LA (2002) Ras participates in the activation of p38 MAPK by interleukin-1 by associating with IRAK, IRAK2, TRAF6, and TAK-1. J Biol Chem 277:7808–7815

Medzhitov R (2001) Toll-like receptors and innate immunity. Nat Rev Immunol 1:135–145

Minns LA, Menard LC, Foureau DM, Darche S, Ronet C, Mielcarz DW, Buzoni-Gatel D, Kasper LH (2006) TLR9 is required for the gut-associated lymphoid tissue response following oral infection of *Toxoplasma gondii*. J Immunol 176:7589–7597

Mishra BB, Gundra UM, Teale JM (2008) Expression and distribution of Toll-like receptors 11–13 in the brain during murine neurocysticercosis. J Neuroinflammation 5:53

Mishra BB, Mishra PK, Teale JM (2006) Expression and distribution of Toll-like receptors in the brain during murine neurocysticercosis. J Neuroimmunol 181:46–56

Mockenhaupt FP, Cramer JP, Hamann L, Stegemann MS, Eckert J, Oh NR, Otchwemah RN, Dietz E, Ehrhardt S, Schroder NW, Bienzle U, Schumann RR (2006) Toll-like receptor (TLR) polymorphisms in African children: common TLR-4 variants predispose to severe malaria. Proc Natl Acad Sci USA 103:177–182

Mun HS, Aosai F, Norose K, Chen M, Piao LX, Takeuchi O, Akira S, Ishikura H, Yano A (2003) TLR2 as an essential molecule for protective immunity against *Toxoplasma gondii* infection. Int Immunol 15:1081–1087

Netea MG, Sutmuller R, Hermann C, Van der Graaf CA, Van der Meer JW, van Krieken JH, Hartung T, Adema G, Kullberg BJ (2004) Toll-like receptor 2 suppresses immunity against *Candida albicans* through induction of IL-10 and regulatory T cells. J Immunol 172: 3712–3718

Nylen S, Sacks D (2007) Interleukin-10 and the pathogenesis of human visceral leishmaniasis. Trends Immunol 28:378–384

Ockenhouse CF, Hu WC, Kester KE, Cummings JF, Stewart A, Heppner DG, Jedlicka AE, Scott AL, Wolfe ND, Vahey M, Burke DS (2006) Common and divergent immune response signaling pathways discovered in peripheral blood mononuclear cell gene expression patterns in presymptomatic and clinically apparent malaria. Infect Immun 74:5561–5573

Parroche P, Lauw FN, Goutagny N, Latz E, Monks BG, Visintin A, Halmen KA, Lamphier M, Olivier M, Bartholomeu DC, Gazzinelli RT, Golenbock DT (2007) Malaria hemozoin is immunologically inert but radically enhances innate responses by presenting malaria DNA to Toll-like receptor 9. Proc Natl Acad Sci USA 104:1919–1924

Pichyangkul S, Yongvanitchit K, Kum-arb U, Hemmi H, Akira S, Krieg AM, Heppner DG, Stewart VA, Hasegawa H, Looareesuwan S, Shanks GD, Miller RS (2004) Malaria blood stage parasites activate human plasmacytoid dendritic cells and murine dendritic cells through a Toll-like receptor 9-dependent pathway. J Immunol 172:4926–4933

Piggott DA, Eisenbarth SC, Xu L, Constant SL, Huleatt JW, Herrick CA, Bottomly K (2005) MyD88-dependent induction of allergic Th2 responses to intranasal antigen. J Clin Invest 115:459–467

Plattner F, Yarovinsky F, Romero S, Didry D, Carlier MF, Sher A, Soldati-Favre D (2008) Toxoplasma profilin is essential for host cell invasion and TLR11-dependent induction of an interleukin-12 response. Cell Host Microbe 3:77–87

Redecke V, Hacker H, Datta SK, Fermin A, Pitha PM, Broide DH, Raz E (2004) Cutting edge:activation of Toll-like receptor 2 induces a Th2 immune response and promotes experimental asthma. J Immunol 172:2739–2743

Reinitz DM, Mansfield JM (1988) Independent regulation of B cell responses to surface and subsurface epitopes of African trypanosome variable surface glycoproteins. J Immunol 141:620–626

Restrepo BI, Alvarez JI, Castano JA, Arias LF, Restrepo M, Trujillo J, Colegial CH, Teale JM (2001) Brain granulomas in neurocysticercosis patients are associated with a Th1 and Th2 profile. Infect Immun 69:4554–4560

Restrepo BI, Llaguno P, Sandoval MA, Enciso JA, Teale JM (1998) Analysis of immune lesions in neurocysticercosis patients: central nervous system response to helminth appears Th1-like instead of Th2. J Neuroimmunol 89:64–72

Rock FL, Hardiman G, Timans JC, Kastelein RA, Bazan JF (1998) A family of human receptors structurally related to *Drosophila* Toll. Proc Natl Acad Sci USA 95:588–593

Ropert C, Franklin BS, Gazzinelli RT (2008) Role of TLRs/MyD88 in host resistance and pathogenesis during protozoan infection: lessons from malaria. Semin Immunopathol 30:41–51

Saint Andre A, Blackwell NM, Hall LR, Hoerauf A, Brattig NW, Volkmann L, Taylor MJ, Ford L, Hise AG, Lass JH, Diaconu E, Pearlman E (2002) The role of endosymbiotic *Wolbachia* bacteria in the pathogenesis of river blindness. Science 295:1892–1895

Sanni LA, Jarra W, Li C, Langhorne J (2004) Cerebral edema and cerebral hemorrhages in interleukin-10-deficient mice infected with *Plasmodium chabaudi*. Infect Immun 72: 3054–3058

Scanga CA, Aliberti J, Jankovic D, Tilloy F, Bennouna S, Denkers EY, Medzhitov R, Sher A (2002) Cutting edge: MyD88 is required for resistance to *Toxoplasma gondii* infection and regulates parasite-induced IL-12 production by dendritic cells. J Immunol 168:5997–6001

Scharton-Kersten TM, Yap G, Magram J, Sher A (1997) Inducible nitric oxide is essential for host control of persistent but not acute infection with the intracellular pathogen *Toxoplasma gondii*. J Exp Med 185:1261–1273

Schleifer KW, Filutowicz H, Schopf LR, Mansfield JM (1993) Characterization of T helper cell responses to the trypanosome variant surface glycoprotein. J Immunol 150:2910–2919

Schofield L, Hackett F (1993) Signal transduction in host cells by a glycosylphosphatidylinositol toxin of malaria parasites. J Exp Med 177:145–153

Schofield L, Hewitt MC, Evans K, Siomos MA, Seeberger PH (2002) Synthetic GPI as a candidate anti-toxic vaccine in a model of malaria. Nature 418:785–789

Schofield L, Novakovic S, Gerold P, Schwarz RT, McConville MJ, Tachado SD (1996) Glycosylphosphatidylinositol toxin of *Plasmodium* up-regulates intercellular adhesion molecule-1, vascular cell adhesion molecule-1, and E-selectin expression in vascular endothelial cells and increases leukocyte and parasite cytoadherence via tyrosine kinase-dependent signal transduction. J Immunol 156:1886–1896

Sternberg JM (2004) Human African trypanosomiasis: clinical presentation and immune response. Parasite Immunol 26:469–476

Takeda K, Akira S (2005) Toll-like receptors in innate immunity. Int Immunol 17:1–14

Takeda K, Kaisho T, Akira S (2003) Toll-like receptors. Annu Rev Immunol 21:335–376

Taylor MJ, Hoerauf A (1999) *Wolbachia* bacteria of filarial nematodes. Parasitol Today 15: 437–442

Theodos CM, Reinitz DM, Mansfield JM (1990) Regulation of B cell responses to the variant surface glycoprotein (VSG) molecule in trypanosomiasis. I. Epitope specificity and idiotypic profile of monoclonal antibodies to the VSG of *Trypanosoma brucei* rhodesiense. J Immunol 144:4011–4021

Thomas PG, Carter MR, Atochina O, Da'Dara AA, Piskorska D, McGuire E, Harn DA (2003) Maturation of dendritic cell 2 phenotype by a helminth glycan uses a Toll-like receptor 4-dependent mechanism. J Immunol 171:5837–5841

Togbe D, Schofield L, Grau GE, Schnyder B, Boissay V, Charron S, Rose S, Beutler B, Quesniaux VF, Ryffel B (2007) Murine cerebral malaria development is independent of toll-like receptor signaling. Am J Pathol 170:1640–1648

Tsan MF (2006) Toll-like receptors, inflammation and cancer. Semin Cancer Biol 16:32–37

van der Heyde HC, Nolan J, Combes V, Gramaglia I, Grau GE (2006) A unified hypothesis for the genesis of cerebral malaria: sequestration, inflammation and hemostasis leading to microcirculatory dysfunction. Trends Parasitol 22:503–508

van der Kleij D, Latz E, Brouwers JF, Kruize YC, Schmitz M, Kurt-Jones EA, Espevik T, de Jong EC, Kapsenberg ML, Golenbock DT, Tielens AG, Yazdanbakhsh M (2002) A novel host-parasite lipid cross-talk. Schistosomal lyso-phosphatidylserine activates toll-like receptor 2 and affects immune polarization. J Biol Chem 277:48122–48129

Vossenkamper A, Struck D, Alvarado-Esquivel C, Went T, Takeda K, Akira S, Pfeffer K, Alber G, Lochner M, Forster I, Liesenfeld O (2004) Both IL-12 and IL-18 contribute to small intestinal Th1-type immunopathology following oral infection with *Toxoplasma gondii*, but IL-12 is dominant over IL-18 in parasite control. Eur J Immunol 34:3197–3207

White AC (1997) Neurocysticercosis: a major cause of neurological disease worldwide. Clin Infect Dis 24:101–113; quiz 114–105

White AC (2000) Neurocysticercosis: updates on epidemiology, pathogenesis, diagnosis, and management. Annu Rev Med 51:187–206

White AC Jr, Robinson P, Kuhn R (1997) *Taenia solium* cysticercosis: host-parasite interactions and the immune response. Chem Immunol 66:209–230

Yano A, Mun HS, Chin M, Norose K, Hata K, Kobayashi M, Aosai F, Iwakura Y (2002) Roles of IFN-gamma on stage conversion of an obligate intracellular protozoan parasite, *Toxoplasma gondii*. Int Rev Immunol 21:405–421

Yap GS, Sher A (1999) Cell-mediated immunity to *Toxoplasma gondii*: initiation, regulation and effector function. Immunobiology 201:240–247

Yarovinsky F, Sher A (2006) Toll-like receptor recognition of *Toxoplasma gondii*. Int J Parasitol 36:255–259

Yarovinsky F, Zhang D, Andersen JF, Bannenberg GL, Serhan CN, Hayden MS, Hieny S, Sutterwala FS, Flavell RA, Ghosh S, Sher A (2005) TLR11 activation of dendritic cells by a protozoan profilin-like protein. Science 308:1626–1629

Zhang D, Zhang G, Hayden MS, Greenblatt MB, Bussey C, Flavell RA, Ghosh S (2004) A toll-like receptor that prevents infection by uropathogenic bacteria. Science 303:1522–1526

Toll-Like Receptors in Neurodegeneration

Trevor Owens

Contents

Abstract Innate pattern recognition receptors are implicated in first-line defense against pathogens but also participate in maintenance of tissue homeostasis and response to injury. This chapter reviews the role of Toll-like receptors (TLRs) in neuronal and glial responses that are associated with neurodegeneration. Accompanying roles for infection and inflammation, involvement in clinical neurodegenerative disorders, and heterogeneity of glial response are discussed. A "strength of signal" hypothesis is advanced in an attempt to reconcile evolutionarily selected and therefore likely beneficial effects of TLR signaling in the nervous system with capability for neurotoxocity and gliotoxicity.

T. Owens
Medical Biotechnology Center, University of Southern Denmark,
Winsloewparken 25, 5000, Odense C, Denmark
e-mail: towens@health.sdu.dk

T. Kielian (ed.), *Toll-like Receptors: Roles in Infection and Neuropathology*,
Current Topics in Microbiology and Immunology 336,
DOI 10.1007/978-3-642-00549-7_6, © Springer-Verlag Berlin Heidelberg 2009

Abbreviations

AD	Alzheimer's disease
CNS	Central nervous system
CpG	Cytosine-phosphate-guanine
dsRNA	Double-stranded RNA
EAE	Experimental autoimmune encephalomyelitis
HIV	Human immunodeficiency virus
HSP	Heat-shock protein
iNOS	Inducible nitric oxide synthase
LPS	Lipopolysaccharide
MyD88	Myeloid differentiation primary response gene (88)
NF-κB	Nuclear factor kappa B
NO	Nitric oxide
NPC	Niemann–Pick type C
SIGIRR	Single immunoglobulin IL-1R-related molecule
STAT	Signal transducer and activator of transcription
TLR	Toll-like receptor

1 Background

Toll was first described in a screen for mutations affecting the development of *Drosophila* (Nusslein-Volhard et al. 1987; Stein et al. 1991). Later it was shown that mutations in Toll reduced fruit fly protection against pathogens (Lemaitre et al. 1996). Mammalian homologs were identified as TLRs in mice and humans (reviewed in Medzhitov 2001). Interest in TLRs was initially primarily focused on pathogen recognition, and they have been shown to exert a central role in the innate immune response, acting as a front-line sentinel system that triggers responses to bacteria, fungi and viruses. Thirteen TLRs have been cloned in humans and mice; most but not all of them are functional, varying between species. This review is intended to provide an update on the involvement of TLRs in neurodegeneration, where TLRs 2, 3 and 4 feature most prominently. A "strength of signal" hypothesis is advanced in an attempt to reconcile evolutionarily selected and therefore likely beneficial effects of TLR signaling in the nervous system with capability for neurotoxocity and gliotoxicity.

Interestingly, TLRs do not appear to play as central a role in mammalian development as *Toll* does in *Drosophila*, since mice lacking individual TLRs can be maintained as viable colonies, at least under pathogen-free conditions. This may simply reflect redundancy between multiple members of an expanded family in vertebrates, although there are also multiple TLRs in *Drosophila*. Whether there are more subtle developmental effects of individual or multiple TLR deficiency remains unanswered at present.

2 Pathogen Recognition Versus Homeostasis

It has become obvious that TLRs do more than recognize molecular pattern epitopes expressed by pathogens ("stranger" signals). Their capacity to recognize endogenous entities points to a possible role in maintenance of homeostasis ("danger" signals) (Gordon 2002; Matzinger 2002). Furthermore, their capacity to induce inflammation, coupled with the growing recognition that inflammation plays a beneficial role in repair following tissue injury, leads to speculation about a beneficial role for TLRs in response to noninfectious tissue injury. In the nervous system, as in other tissues, there are multiple associations between inflammation and degeneration. To list a few, ischemia-induced neurodegeneration in clinical stroke is exacerbated by leukocyte infiltration and the subsequent immune and glial inflammatory response; immune infiltration and immune + glial inflammation in multiple sclerosis are associated with both demyelination and axonal damage; and neuronal loss in Alzheimer's disease (AD) is accompanied by glial reactivity. Thus, it can be presumed that signals that induce or regulate central nervous system (CNS) inflammation are candidate signals for neurodegeneration. Given the role of TLRs in inducing inflammatory responses, and their ability to respond to endogenous ligands, it is consistent that they should be implicated in neurodegeneration and neurodegenerative disease (Kielian 2006).

3 Inflammation and Degeneration: Immune Concepts

TLRs are well known for their capacity to induce protective responses against pathogens. These responses include cellular cytotoxicity, a necessary component of immune defense. Macrophages play a central role in the immune response to pathogens. This includes the capacity of activated or inflamed macrophages to kill bacteria and also host cells infected with bacteria, mycoplasma or viruses, through mechanisms that include direct killing via reactive oxygen and nitrogen species or cytotoxic cytokines, as well as more indirect mechanisms in collaboration with antibodies, complement and other effector cells. These cytopathic actions of macrophages can all be triggered by TLR signaling. An evolutionarily inherent risk of placing TLRs in a critical directorial role in innate responses in the CNS is that the cytopathic arm of their downstream effects may dominate the outcome. Microglia are CNS-resident cells that share a bone marrow origin and effector capability with macrophages and indeed are considered "brain macrophages." It is consistent with the lineage and phenotypic relationships between microglia and macrophages, therefore, that the glial cell most associated with neurodegeneration should be the microglial cell. To date, almost all reported instances where TLRs are implicated in neurodegeneration involve microglial activation, and most of these involve TLR2 or TLR4 signaling (Table 1).

Table 1 TLRs in neurodegeneration: summary of principal findings

TLR	Pro- or antidegenerative outcome	Cell type	TLR ligand/agonist	Mediators or mechanism	References
2	Wallerian degeneration	Schwann cells	Necrotic neurons	NO, TNFα	Lee et al. (2006)
	Wallerian degeneration	Microglia	Zymosan	NO, CCL2	Boivin et al. (2007)
	Microglial/leukocyte mobilization	Microglia	Unknown	CCL2, TNFα	Babcock et al. (2006)
	Recovery from SC injury	Microglia	Unknown	IL-1β, NO	Kigerl et al. (2007)
	Pain hypersensitivity	Microglia	Unknown	NO, TNFα, IL-1β, IL-6	Kim et al. (2007)
	OL killing/NDG	Microglia	Group B *Streptococcus*	NO release	Lehnardt et al. (2006)
	Microglial apoptosis	Microglia	Group B *Streptococcus*	Caspase 8	Lehnardt et al. (2007b)
	Neuronal killing	Microglia	PAM3CysSK4	Not described	Hoffmann et al. (2007)
3	Wallerian degeneration	Schwann cells	Necrotic neurons	NO, TNFα	Lee et al. (2006)
	Wallerian degeneration	Microglia	Poly I:C	NO, CCL2	Boivin et al. (2007)
	Demyelination/NDG	Schwann cells	dsRNA, poly I:C	iNOS/NO	Lee et al. (2007)
4	Wallerian degeneration	Microglia	LPS	NO, CCL2	Boivin et al. (2007)
	Wallerian degeneration	Schwann cells	Necrotic neurons	CCL2	Karanth et al. (2006)
	Recovery from SC injury	Microglia	Unknown	IL-1β, NO	Kigerl et al. (2007)
	OL killing/NDG	Microglia	LPS	Not described	Lehnardt et al. (2002), Lehnardt et al. (2003)
	Neuron killing	Microglia	HSP60	NO	Lehnardt et al. (2008)
	Unknown	Microglia	Fibronectin EDA	NO, TNFα	Goos et al. (2007)
	DA neuron loss	Microglia	LPS	Superoxide	Ling et al. (2006)
	Microglial apoptosis	Microglia	LPS	IFNβ, Caspase 3, 11	Ji et al. (2007), Jung et al. (2005)
8	Neuronal apoptosis	Neurons	Resiquimod	Caspase(s)	Ma et al. (2006)
9	Glioma apoptosis	GL261 glioma	CpG	Caspase 8	El Andaloussi et al. (2006)

DA, dopaminergic; *OL*, oligodendrocyte; *NDG*, neurodegeneration; *SC*, spinal cord

One issue to be considered when interpreting the physiological role of TLR signaling is that of agonist ligands, as exemplified by TLR4 signaling. The classical TLR4 agonist is bacterial lipopolysaccharide (LPS). Although it is recognized by TLR4, the LPS-binding co-receptor CD14 and the non-LPS-binding molecule MD2 are also essential for signaling (Akira and Takeda 2004). Furthermore, TLR4 can signal via both myeloid differentiation primary response gene 88 (MyD88)-dependent and -independent pathways. However, TLR4 signals can operate in the absence of bacterial infection and many other TLR4 agonists have been described, including free fatty acids, hyaluronin, fibronectin, and heat-shock proteins, and others that remain to be identified (Larsen et al. 2007). Determining that a biological response is TLR4 dependent is only a first step, since TLR4 signaling itself depends on a complex of co-receptors, and signaling outcomes may vary with the nature of the agonist or ligand. Therefore, distinct outcomes of TLR4 and other TLR signaling pathways may reflect the operation of different agonists or ligands. Analogously, TLR2 signaling is influenced by whether it dimerizes with TLR1 or TLR6 (Akira and Takeda 2004).

4 Beneficial Outcomes of TLR Signaling in Microglia

4.1 Wallerian Degeneration

Activated glial cells are implicated in Wallerian degeneration that occurs in the spinal cord after peripheral nerve injury. Despite the term "degeneration," this process is ultimately beneficial, leading to clearance of debris and restoration of microenvironmental homeostasis, both of which can promote repair and regeneration. The degenerative component is thus postinjury. The predominant cellular mechanism in Wallerian degeneration is phagocytosis, which is classically triggered by TLR signaling. Sciatic nerve Schwann cells induce Wallerian degeneration via the production of mediators such as nitric oxide (NO) and the chemokine CCL2. There is much interest in the receptor(s) involved in Schwann cell recognition of necrotic neurons, and the use of gene-deficient mice (Lee et al. 2006) or specific antibodies (Karanth et al. 2006) has shown roles for TLR2 and TLR3, and TLR4, respectively. In another study, mice defective in the expression of TLR2 or TLR4 showed deficiencies in Wallerian degeneration in the sciatic nerve distal stump after sciatic nerve transection (Boivin et al. 2007). Microglial activation and phagocytic response could be experimentally triggered by in vivo administration of TLR ligands to the site of sciatic nerve lesion.

4.2 Nerve Injury

Such a beneficial microglial response is analogous to that reported by Kigerl et al., who noted that mice lacking expression of TLR2 or TLR4 showed impaired functional recovery from experimental spinal cord injury (Kigerl et al. 2007). The correlation with microglial activation was compelling given the fulminant microgliosis that accompanied percussive injury, although a demonstration such as that by Boivin et al. that TLR signaling could subserve a physiological role was not made. Other studies, reviewed elsewhere (Larsen et al. 2007), suggest roles for microglial TLR2 signals that can also be interpreted as protective in the induction of pain hypersensitivity after spinal nerve transection, and the regulation of microgliosis and leukocyte infiltration at sites of synaptic degeneration (Babcock et al. 2006; Kim et al. 2007). In both of these studies, endogenous ligands were presumed but not identified, and roles for glia-derived cytokines and chemokines were shown (Table 1).

5 TLR-Driven Neurotoxicity

5.1 Neuronal Injury

The ability of microglia to kill neurons via the action of reactive nitrogen species such as NO has been described by numerous studies. A role for TLR4 in driving neurotoxic NO was described by Lehnardt et al., who also showed that this could be provoked by the TLR4 agonist heat-shock protein 60 (HSP60). HSP60 is released by damaged or dying neurons, and this triggered microglial responses in vitro (Lehnardt et al. 2008). It is assumed that similar effects occur in vivo, and this may be a mechanism for the microglial reactivity that is associated with any instance of neuronal death or damage (Hanisch and Kettenmann 2007). These in vitro findings suggest that HSP60 released by dying neurons acts as a feedforward signal to provoke further neurotoxicity. HSP60-triggered NO may represent a mechanism that is designed to ensure the removal of damaged or dysfunctional neurons in vivo. This appears distinct from proregenerative roles for TLR4 signaling seen in other studies (Kigerl et al. 2007). Whether these apparently distinct TLR4-signaled actions reflect a specific effect of HSP60, released under select circumstances, or dysregulation or amplification of an HP60-driven physiological response during in vitro studies remains to be determined.

Complementary observations have been made for alternately spliced cellular fibronectin, which is associated with tissue remodeling during inflammation. The Type III repeat extra domain A of fibronectin is an agonist ligand for TLR4 (Okamura et al. 2001). Fibronectin levels are elevated in the cerebrospinal fluid of patients with meningitis and can enhance TLR4-driven activation of mouse microglia (Goos et al. 2007), which creates a plausible mechanism for microglial activation and downstream effects in human brain infection and inflammation. As-yet molecularly uncharacterized factors released by neurons that modulate TLR4 responses of microglia or Schwann cells may have analogous roles in vivo (Karanth et al. 2006; Holm, unpublished).

5.2 *Ischemia*

A prodegenerative role for TLR signaling has also been suggested by studies in ischemia. Hippocampal neurodegeneration follows global cerebral ischemia and reperfusion. Activation of an inflammatory response is a critical event in this cascade, involving infiltrating leukocytes and glial cell activation. The role of TLR signals in activating the inflammasome, which controls caspase activation and the processing of substrates such as IL-1β, is of interest (Trendelenburg 2008). TLR4-deficient mice showed decreased neuronal loss and apoptosis following global cerebral ischemia and reperfusion, with concomitant decreased production of inflammatory cytokines and TLR4-associated signaling (Hua et al. 2007). Although these findings remain correlative, and the exact mechanism by which TLR4 deficiency influences ischemia-induced neuronal loss was not determined, the study adds to our understanding of TLR4 as a critical innate signal in brain response to injury. Emphasizing the role of innate signals in glial responses, it has been shown in two studies that mice lacking TLR2 (normally expressed by lesion-associated microglia but also expressed by astrocytes) also show decreased infarction and neuronal loss in response to focal cerebral ischemia (Lehnardt et al. 2007a; Ziegler et al. 2007).

5.3 *Neurodegenerative Disorders*

Analogously, increased TLR4 expression and activation of microglia with inducible nitric oxide synthase (iNOS) expression in the substantia nigra pars compacta accompanied multisystem atrophy and loss of dopaminergic neurons in transgenic mice with oligodendroglial α-synuclein inclusions (Stefanova et al. 2007). Again, the correlation with TLR4 expression is suggestive, although a role for TLR4 must be considered speculative in the absence of intervention studies. A possible mechanism whereby TLR4 signaling to microglia could mediate loss of dopaminergic neurons is the production of superoxide, which is induced by LPS, and which selectively kills dopaminergic neurons in vitro and in vivo (Qin et al. 2004; Ling et al. 2006). As with NO production, the innate receptor system here acts blindly, and the cytopathic consequences of TLR signaling may be considered inadvertent and unrelated to the evolutionary pressures that led to their existence in the CNS in the first place.

Niemann-Pick disease involves progressive neurodegeneration starting in childhood, and is related to dysfunctional endosomal or lysosomal genes with resulting lipid storage disorder in many tissues. Glial activation in the CNS of Niemann–Pick Type C (NPC) patients has been proposed to be causative, and increased expression of TNF-α, IL-6 and IL-1β as well as deregulated nuclear factor kappa B (NF-κB) by astrocytes and neurons has been reported (Bi et al. 2005; Suzuki et al. 2007). A recent study of NPC fibroblasts revealed overexpression and endosomal accumulation of TLR4, and IL-6 production was decreased in fibroblasts from TLR4-deficient mice (Suzuki et al. 2007). It has been reported that macrophages of AD patients do not transport amyloid-β protein into endosomes as effectively as controls, and that this correlates with depressed transcription of a number of TLR genes, including TLRs 3, 4, 5, 7, 8, 9 and 10 (Fiala et al. 2007). It has been proposed that downstream signaling from TLR4 may be a common feature of NPC and AD (Suzuki et al. 2007). This raises questions about the nature of TLR4 agonists/ligands, but is generally consistent with the theme that TLR-driven glial activation plays a critical role in neurodegenerative disorders. Whether and to what extent this would contribute to the impact of infection on neurodegenerative disease is another interesting question, and one that has received fuller treatment elsewhere (Perry et al. 2007).

5.4 *Infection*

The emphasis in many studies on endogenous ligands or agonists, and roles for TLRs that are independent of infection, reflects that interest is principally in sterile injury. Although other chapters in this volume will deal at greater length with pathogen-induced glial response, infectious pathogens of course may themselves provoke glial responses that have neurodegenerative impact (Kielian 2006; Perry et al. 2007). Group B streptococci are a source of TLR2 agonist activity that, analogous to the synthetic TLR2 agonist PAM3CysSK4 as well as LPS and other TLR4 agonists that have been discussed, can provoke neurotoxin release by microglia in vitro. Intrathecal administration of PAM3CysSK4 to mice induced cerebrospinal fluid pleocytosis and neuronal apoptosis (Hoffmann et al. 2007) and TLR2-triggered microglia produced neurocytopathic supernatants in vitro, which complements and identifies an experimental mechanism for the TLR2/MyD88-dependent and NO-mediated meningitis that is induced by Group B streptococci (Lehnardt et al. 2006).

5.5 *TLR-Driven Gliotoxicity*

Not only do TLR-signaled microglia exert neurotoxic effects, but they can themselves be induced to commit suicide. Thus, LPS has been reported to induce apoptosis of microglia in vitro (Jung et al. 2005; Ji et al. 2007). The mechanism was shown in one study to involve induction of interferon regulatory factor 3 and interferon beta, which then induced the executioners caspase 3 and caspase 11 (Jung et al. 2005). Interestingly, in this study the TLR2 ligand/agonists lipoteichoic acid, peptidoglycan or PAM3CysSK4 did not induce apoptosis in microglia. In contrast to that finding, Group B streptococci were reported to induce microglial apoptosis via TLR2 by Lehnardt et al. (2007b), and this was due to the activation of another executioner caspase, caspase 8, but not caspase 3. Reasons for these discrepant findings are not clear, but it is possible that a bacterial cell-associated TLR2 ligand might signal differently than a soluble ligand.

6 Strength of Signal: A Hypothesis

An aspect that requires further consideration is whether all microglial activation is neurotoxic or whether this depends on the circumstances. Many in vitro studies have shown that activated microglia release neurotoxic NO. Nevertheless, many observations in situ and in vivo show activated microglia that are not apparently as detrimental to surrounding neurons as can be achieved in vitro; for instance, the lack of degeneration in response to LPS under physiological conditions of glucocorticoid regulation (Nadeau and Rivest 2003). Resolution of these discrepancies probably lies in the realization that although microglial activation universally results in phenotypic changes (upregulation of CD11b/Mac-1, morphologic changes), the biological outcome may vary. The concept of protective versus detrimental macrophage activation (Gordon 2003) can be extended to microglia, and the quality of the cytokine milieu significantly influences microglial responses. It is possible that in vitro studies selectively mimic proinflammatory or degenerative milieu, whereas in vivo manipulations promote more benign or even protective responses. The possibility that in vitro systems allow responses or levels of response that are unattainable in vivo is consistent with a speculative hypothesis that the principal factor in determining outcome of glial response is strength of signal (Fig. 1). In vitro this can be controlled via the concentration of the TLR ligand/agonist, as has been shown indirectly by comparing studies using high and low doses of LPS in inducing astrocyte responses in vitro (Carpentier et al. 2005; Holm, unpublished, vide infra). It is more difficult to control agonist dosages in vivo, but it may be relevant that in subclinical injury responses the outcome of TLR ligation was nondegenerative and indeed was restricted to one or a few TLR signals (Babcock et al. 2006), in contrast

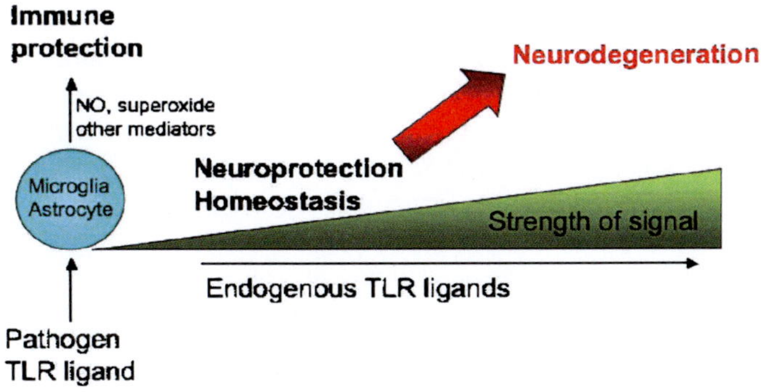

Fig. 1 Schematic to illustrate a strength of signal hypothesis and the role of TLRs in neurodegeneration, as discussed in this review

to responses to more traumatic injuries (Kigerl et al. 2007; Babcock et al. 2008). However, this speculation is based on a small number of studies and remains to be definitively tested.

One approach for resolving issues related to physiological relevance is to apply TLR stimuli in vivo and directly assess responses, as was used by Hoffmann et al. (2007) and by Rivest and colleagues in their studies of glucocorticoid regulation of inflammation-associated neurodegeneration (Nadeau and Rivest 2003) and Wallerian degeneration (Boivin et al. 2007). A similar approach was used by Lee et al. in studies of TLR3 signaling to Schwann cells in the rat sciatic nerve (Lee et al. 2007). Other studies from the same group had shown that Schwann cells, which can induce peripheral nerve Wallerian degeneration, upregulate iNOS expression in a TLR2- and TLR3-dependent manner in response to necrotic neuronal cells (Lee et al. 2006). Injection of double-stranded RNA (dsRNA) to the rat sciatic nerve induced iNOS gene expression and demyelination, and although the exact cellular target of dsRNA was not established, the findings are consistent with in vitro studies showing that Schwann cells express TLR3 and that its triggering induced iNOS expression and death of co-cultured neurons (Lee et al. 2007). In this case, the in vivo outcome was more tempered than the in vitro effects in that neurodegeneration was not described, consistent with the commentary above for microglia. However, these in vivo studies collectively support a cytopathic role for TLR-activated glial cells in vivo.

7 Clinical Correlates

Another approach is to inquire whether TLRs are implicated in clinical neurodegenerative diseases. Letiembre et al. screened for expression of TLRs and CD14 and found that although TLR2 and CD14 were both upregulated in autopsy material from AD and in etiologically diverse animal models of AD, they were not associated with Lewy body-associated dementia or amyotrophic lateral sclerosis. TLR2 and CD14 were co-localized to microglia in AD and the animal models, whereas other TLRs showed variable patterns of upregulation. Although TLR4 was not upregulated in any of the mouse models for AD, this may plausibly reflect a balance between upregulation and ligand-induced downregulation (Letiembre et al. 2007). Another study correlated a Asp299Gly TLR4 polymorphism with susceptibility to late-onset AD, in that the frequency of this polymorphism, which is associated with attenuated inflammatory responses, was significantly higher in controls than in the AD cohort (Minoretti et al. 2006). These findings are broadly consistent with other studies that have shown elevated expression of individual or selected TLRs in AD and multiple sclerosis (Bsibsi et al. 2002; Fassbender et al. 2004). Salaria et al. carried out analogous screens for signaling pathways in human immunodeficiency virus (HIV)-associated neurodegeneration and also identified TLR-related pathways (Salaria et al. 2007). There was a statistically significant association of elevated transcripts for nine TLRs or TLR-related genes, including TLR3, TLR4 and IκB-β, coincident with a reduction in dendritic and synaptic staining, and quantitative real-time RT-PCR confirmed an association with TLR4. Whereas TLR4 expression correlated positively with indices of neurodegeneration, expression of single immunoglobulin IL-1R-related molecule (SIGIRR), which counteracts TLR4 signaling, correlated negatively, consistent overall with a neurodegenerative role for TLR4 (Salaria et al. 2007). While such studies may suggest a neuroinflammatory component to these neurodegenerative diseases, they do not discriminate between detrimental and protective roles for the TLR2- or CD14-expressing microglia.

8 TLR2/4 Expression by Astrocytes

Many of the above-described studies describe upregulation and signaling through TLR2 on CNS microglia. There is a preponderance of studies identifying this TLR within the CNS, partly reflecting the availability of reagents, but also perhaps suggesting a role in the recognition of endogenous ligand(s). Also notable, and especially highlighted in the study by Babcock et al. (Babcock et al. 2006), was the complete absence of induction of TLR2 by astrocytes, despite their side-by-side proximity to TLR2-positive microglia. This suggests exquisite specificity of the inducing agent(s) that act on microglia. Attempts to dissect this specificity in vitro have generated mixed interpretations, with many studies showing that astrocytes can be induced to express TLR2 and respond to TLR2 ligands. In some cases the inducing agent has been the TLR4 ligand LPS, and this might not have acted directly on astrocytes. Although HIV gp120 has been reported to induce TLR4 expression by human astrocytes (Salaria et al. 2007), and cultured murine astrocytes are reported to express TLR4 mRNA (Bowman et al. 2003), other reports have failed to identify TLR4 or CD14 on human or rodent astrocytes (Beschorner et al. 2002a,b; Lehnardt et al. 2002; Farina et al. 2007). Thus it has been suggested that the apparent astrocyte response to TLR4 ligands is in fact an indirect response to TLR4-activated microglia that produce cytokines (Kielian 2006; Saura 2007), and so it is microglial cytokines such as IL-1β that induce TLR2 on astrocytes. This leads to further questions about the quality of the microglia-generated cytokine milieu in vivo, in which astrocytes failed to induce TLR2—not least the fact that one of the prominent cytokines induced by axonal lesioning was IL-1β (Babcock et al. 2006).

8.1 Astrocytes in Neurodegeneration

Although classically regarded as a beneficial or protective cell type (Farina et al. 2007), a role for astrocytes in neurotoxicity was suggested by an elegant study in which the transcription factor NF-κB was selectively disabled in astrocytes in transgenic mice expressing a dominant-negative IkB-α construct under the control of an astrocyte-specific glial fibrillary acidic protein promoter (Brambilla et al. 2005). These animals showed accelerated recovery from spinal cord injury, suggesting that NF-κB-dependent astrocyte responses were detrimental in wild-type mice. NF-κB is a point of convergence for TLR signals (Akira and Takeda 2004), although TLR involvement in the spinal cord injury experiments in this study was not demonstrated. Our own studies show that the signal transducer and activator of transcription 2 (STAT2) was selectively upregulated and phosphorylated in astrocytes in the zone of synaptic degeneration in the dentate gyrus after entorhinal axotomy, and that too was NF-κB dependent (Khorooshi et al. 2008). In considering whether a TLR signal would be involved in the astrocyte response, it is telling that a frequently reported TLR on astrocytes is TLR3 (Farina et al. 2007), which recognizes double-stranded RNA and so acts as an innate immune receptor for virus infection. However, TLR3 is unique among TLRs in not having an MyD88-dependent signaling pathway (Akira and Takeda 2004), and it has separately been shown that the glial response to axonal injury in the entorhinal lesion model is largely MyD88 dependent (Babcock et al. 2008). This is a complex area and the data to hand at time of writing are insufficient to allow firm conclusions as to the role of TLRs in astroglial responses to neuronal injury.

9 Other TLRs

There are isolated reports describing the roles of TLRs other than TLRs 2, 3, and 4 in neurodegeneration. The depressed transcription of TLRs 5, 7, 8, 9 and 10 in AD macrophages has already been discussed (Fiala et al. 2007). This is a relatively young field, and it is likely that that the evidence for broader involvement of TLRs in neurodegenerative processes will continue to grow.

Vartanian and colleagues reported that TLR8 plays a role in suppressing neuronal outgrowth, and showed that its triggering by a synthetic ligand could induce apoptosis of neurons in culture (Ma et al. 2006). This observation differs in an important respect from those described above, in that apoptosis-signaling TLR8 was itself expressed by the susceptible neurons, rather than acting via a glial intermediate. Since TLR8 is strongly expressed early in mouse development, it is tempting to speculate that this inhibitory effect plays a role in shaping the developing nervous system (Larsen et al. 2007). Neuronal growth inhibition by TLR3 signaling has also been reported but did not appear to induce apoptosis (Cameron et al. 2007).

Induction of the demyelinating disease experimental autoimmune encephalomyelitis (EAE) was shown to be dependent on TLR9 (Prinz et al. 2006). This was also shown in adoptive transfer experiments transferring myelin-specific wild-type T cells to TLR9-deficient mice, so that potentially cytosine-phosphate-guanine (CpG)-containing adjuvants could not confound the interpretation. The findings indicate a role for TLR9 expressed on a non-T cell, possibly within the CNS, responding to an endogenous ligand. Since one of the pathological consequences in EAE is axonal damage, this constitutes evidence—albeit indirect—for a role for

TLR9 in neurodegeneration. A conflicting finding, that mice lacking TLR4 or TLR9 showed exacerbated EAE (Marta et al. 2008), is difficult to reconcile with the findings from Prinz et al., although both studies showed that MyD88-deficient mice are EAE resistant, and it can be argued that both studies did in fact show a role for TLR9 in EAE, but that differences in immunization strategy obscured regulatory effects in the studies by Prinz et al. (Marta et al. 2008). Resolution of this issue must await further analysis.

TLR9 is overexpressed in human and mouse glioma lines and has been implicated in glioma killing. Stimulation with agonist/ligand CpG dinucleotide-containing oligonucleotides induced TLR9 downregulation and apoptosis of GL261 glioma cells in vitro and in vivo (El Andaloussi et al. 2006). As with TLR2-mediated microglial apoptosis, the mechanism of killing involved caspase 3 activation (El Andaloussi et al. 2006). It is interesting to consider whether endogenous ligands might subserve a protective role against such tumors, and of course the potential to exploit this in therapy is noted.

10 Conclusions

Links between inflammation and degeneration in the nervous system are increasingly being appreciated. It is difficult to identify situations where either of these two "edges of a sword" can be shown in the absence of the other. The central role of innate pattern-recognition receptors in the initiation of inflammatory responses makes it almost self-evident that TLRs would play a significant role in neurodegenerative processes and disorders. Of interest is the prominent role played by TLR-expressing glial cells as intermediates in clinical and experimental neurodegeneration. This may reflect an immunologically oriented bias towards the expression of receptors guiding inflammation on cells that can present antigen and produce cytokines, although the possibility that such cells contribute to an archetypal role in development cannot be excluded. Certainly it can be assumed that evolutionary selection did not place TLR-based signaling mechanisms in the CNS purely to cause neurodegeneration, so one concludes that the potential for TLR signaling to initiate neurodegeneration reflects dysregulation of an otherwise benign process.

Acknowledgments Research in my laboratory is supported by Scleroseforeningen, Novo Nordisk Fonden, Lundbeckfonden and The Danish Agency for Research and Innovation. I acknowledge the EU/ESF COST Action BM0603 "Neurinfnet" for its support of collaborative interactions that contributed to this review.

References

Akira S, Takeda K (2004) Toll-like receptor signaling. Nat Rev Immunol 4:499–511

Babcock AA, Toft-Hansen H, Owens T (2008) Signaling through MyD88 regulates leukocyte recruitment after brain injury. J Immunol 181:6481–6490

Babcock AA, Wirenfeldt M, Holm T, Nielsen HH, Dissing-Olesen L, Toft-Hansen H, Millward JM, Landmann R, Rivest S, Finsen B, Owens T (2006) Toll-like receptor 2 signaling in response to brain injury: an innate bridge to neuroinflammation. J Neurosci 26:12826–12837

Beschorner R, Nguyen TD, Gozalan F, Pedal I, Mattern R, Schluesener HJ, Meyermann R, Schwab JM (2002a) CD14 expression by activated parenchymal microglia/macrophages and infiltrating monocytes following human traumatic brain injury. Acta Neuropathol 103:541–549

Beschorner R, Schluesener HJ, Gozalan F, Meyermann R, Schwab JM (2002b) Infiltrating CD14+ monocytes and expression of CD14 by activated parenchymal microglia/macrophages contribute to the pool of CD14+ cells in ischemic brain lesions. J Neuroimmunol 126:107–115

Bi X, Liu J, Yao Y, Baudry M, Lynch G (2005) Deregulation of the phosphatidylinositol-3 kinase signaling cascade is associated with neurodegeneration in Npc1–/– mouse brain. Am J Pathol 167:1081–1092

Boivin A, Pineau I, Barrette B, Filali M, Vallieres N, Rivest S, Lacroix S (2007) Toll-like receptor signaling is critical for Wallerian degeneration and functional recovery after peripheral nerve injury. J Neurosci 27:12565–12576

Bowman CC, Rasley A, Tranguch SL, Marriott I (2003) Cultured astrocytes express toll-like receptors for bacterial products. Glia 43:281–291

Brambilla R, Bracchi-Ricard V, Hu WH, Frydel B, Bramwell A, Karmally S, Green EJ, Bethea JR (2005) Inhibition of astroglial nuclear factor kappaB reduces inflammation and improves functional recovery after spinal cord injury. J Exp Med 202:145–156

Bsibsi M, Ravid R, Gveric D, van Noort JM (2002) Broad expression of Toll-like receptors in the human central nervous system. J Neuropathol Exp Neurol 61:1013–1021

Cameron JS, Alexopoulou L, Sloane JA, DiBernardo AB, Ma Y, Kosaras B, Flavell R, Strittmatter SM, Volpe J, Sidman R, Vartanian T (2007) Toll-like receptor 3 is a potent negative regulator of axonal growth in mammals. J Neurosci 27:13033–13041

Carpentier PA, Begolka WS, Olson JK, Elhofy A, Karpus WJ, Miller SD (2005) Differential activation of astrocytes by innate and adaptive immune stimuli. Glia 49:360–374

El Andaloussi A, Sonabend AM, Han Y, Lesniak MS (2006) Stimulation of TLR9 with CpG ODN enhances apoptosis of glioma and prolongs the survival of mice with experimental brain tumors. Glia 54:526–535

Farina C, Aloisi F, Meinl E (2007) Astrocytes are active players in cerebral innate immunity. Trends Immunol 28:138–145

Fassbender K, Walter S, Kuhl S, Landmann R, Ishii K, Bertsch T, Stalder AK, Muehlhauser F, Liu Y, Ulmer AJ, Rivest S, Lentschat A, Gulbins E, Jucker M, Staufenbiel M, Brechtel K, Walter J, Multhaup G, Penke B, Adachi Y, Hartmann T, Beyreuther K (2004) The LPS receptor (CD14) links innate immunity with Alzheimer's disease. FASEB J 18:203–205

Fiala M, Liu PT, Espinosa-Jeffrey A, Rosenthal MJ, Bernard G, Ringman JM, Sayre J, Zhang L, Zaghi J, Dejbakhsh S, Chiang B, Hui J, Mahanian M, Baghaee A, Hong P, Cashman J (2007) Innate immunity and transcription of MGAT-III and Toll-like receptors in Alzheimer's disease patients are improved by bisdemethoxycurcumin. Proc Nat Acad Sci USA 104:12849–12854

Goos M, Lange P, Hanisch UK, Prinz M, Scheffel J, Bergmann R, Ebert S, Nau R (2007) Fibronectin is elevated in the cerebrospinal fluid of patients suffering from bacterial meningitis and enhances inflammation caused by bacterial products in primary mouse microglial cell cultures. J Neurochem 102:2049–2060

Gordon S (2002) Pattern recognition receptors: doubling up for the innate immune response. Cell 111:927–930

Gordon S (2003) Alternative activation of macrophages. Nat Rev Immunol 3:33–35

Hanisch UK, Kettenmann H (2007) Microglia: active sensor and versatile effector cells in the normal and pathologic brain. Nat Neurosci 10:1387–1394

Hoffmann O, Braun JS, Becker D, Halle A, Freyer D, Dagand E, Lehnardt S, Weber JR (2007) TLR2 mediates neuroinflammation and neuronal damage. J Immunol 178:6476–6481

Hua F, Ma J, Ha T, Xia Y, Kelley J, Williams DL, Kao RL, Browder IW, Schweitzer JB, Kalbfleisch JH, Li C (2007) Activation of Toll-like receptor 4 signaling contributes to hippocampal neuronal death following global cerebral ischemia/reperfusion. J Neuroimmunol 190:101–111

Ji KA, Yang MS, Jeong HK, Min KJ, Kang SH, Jou I, Joe EH (2007) Resident microglia die and infiltrated neutrophils and monocytes become major inflammatory cells in lipopolysaccharide-injected brain. Glia 55:1577–1588

Jung DY, Lee H, Jung BY, Ock J, Lee MS, Lee WH, Suk K (2005) TLR4, but not TLR2, signals autoregulatory apoptosis of cultured microglia: a critical role of IFN-beta as a decision maker. J Immunol 174:6467–6476

Karanth S, Yang G, Yeh J, Richardson PM (2006) Nature of signals that initiate the immune response during Wallerian degeneration of peripheral nerves. Exp Neurol 202:161–166

Khorooshi R, Babcock AA, Owens T (2008) NF-kappaB-driven STAT2 and CCL2 expression in astrocytes in response to brain injury. J Immunol 181:7284–7291

Kielian T (2006) Toll-like receptors in central nervous system glial inflammation and homeostasis. J Neurosci Res 83:711–730

Kigerl KA, Lai W, Rivest S, Hart RP, Satoskar AR, Popovich PG (2007) Toll-like receptor (TLR)-2 and TLR-4 regulate inflammation, gliosis, and myelin sparing after spinal cord injury. J Neurochem 102:37–50

Kim D, Kim MA, Cho IH, Kim MS, Lee S, Jo EK, Choi SY, Park K, Kim JS, Akira S, Na HS, Oh SB, Lee SJ (2007) A critical role of toll-like receptor 2 in nerve injury-induced spinal cord glial cell activation and pain hypersensitivity. J Biol Chem 282:14975–14983

Larsen PH, Holm TH, Owens T (2007) Toll-like receptors in brain development and homeostasis. Sci STKE 2007:pe47

Lee H, Jo EK, Choi SY, Oh SB, Park K, Kim JS, Lee SJ (2006) Necrotic neuronal cells induce inflammatory Schwann cell activation via TLR2 and TLR3: implication in Wallerian degeneration. Biochem Biophys Res Commun 350:742–747

Lee H, Park C, Cho IH, Kim HY, Jo EK, Lee S, Kho HS, Choi SY, Oh SB, Park K, Kim JS, Lee SJ (2007) Double-stranded RNA induces iNOS gene expression in Schwann cells, sensory neuronal death, and peripheral nerve demyelination. Glia 55:712–722

Lehnardt S, Henneke P, Lien E, Kasper DL, Volpe JJ, Bechmann I, Nitsch R, Weber JR, Golenbock DT, Vartanian T (2006) A mechanism for neurodegeneration induced by group B streptococci through activation of the TLR2/MyD88 pathway in microglia. J Immunol 177:583–592

Lehnardt S, Lachance C, Patrizi S, Lefebvre S, Follett PL, Jensen FE, Rosenberg PA, Volpe JJ, Vartanian T (2002) The toll-like receptor TLR4 is necessary for lipopolysaccharide-induced oligodendrocyte injury in the CNS. J Neurosci 22:2478–2486

Lehnardt S, Lehmann S, Kaul D, Tschimmel K, Hoffmann O, Cho S, Krueger C, Nitsch R, Meisel A, Weber JR (2007a) Toll-like receptor 2 mediates CNS injury in focal cerebral ischemia. J Neuroimmunol 190:28–33

Lehnardt S, Massillon L, Follett P, Jensen FE, Ratan R, Rosenberg PA, Volpe JJ, Vartanian T (2003) Activation of innate immunity in the CNS triggers neurodegeneration through a Toll-like receptor 4-dependent pathway. Proc Natl Acad Sci U S A 100:8514–8519

Lehnardt S, Schott E, Trimbuch T, Laubisch D, Krueger C, Wulczyn G, Nitsch R, Weber JR (2008) A vicious cycle involving release of heat shock protein 60 from injured cells and activation of toll-like receptor 4 mediates neurodegeneration in the CNS. J Neurosci 28:2320–2331

Lehnardt S, Wennekamp J, Freyer D, Liedtke C, Krueger C, Nitsch R, Bechmann I, Weber JR, Henneke P (2007b) TLR2 and caspase-8 are essential for group B *Streptococcus*-induced apoptosis in microglia. J Immunol 179:6134–6143

Lemaitre B, Nicolas E, Michaut L, Reichhart JM, Hoffmann JA (1996) The dorsoventral regulatory gene cassette spatzle/Toll/cactus controls the potent antifungal response in *Drosophila* adults. Cell 86:973–983

Letiembre M, Liu Y, Walter S, Hao W, Pfander T, Wrede A, Schulz-Schaeffer W, Fassbender K (2007) Screening of innate immune receptors in neurodegenerative diseases: a similar pattern. Neurobiol Aging 30:759–768

Ling Z, Zhu Y, Tong C, Snyder JA, Lipton JW, Carvey PM (2006) Progressive dopamine neuron loss following supra-nigral lipopolysaccharide (LPS) infusion into rats exposed to LPS prenatally. Exp Neurol 199:499–512

Ma Y, Li J, Chiu I, Wang Y, Sloane JA, Lu J, Kosaras B, Sidman RL, Volpe JJ, Vartanian T (2006) Toll-like receptor 8 functions as a negative regulator of neurite outgrowth and inducer of neuronal apoptosis. J Cell Biol 175:209–215

Marta M, Andersson A, Isaksson M, Kampe O, Lobell A (2008) Unexpected regulatory roles of TLR4 and TLR9 in experimental autoimmune encephalomyelitis. Eur J Immunol 38: 565–575

Matzinger P (2002) The danger model: a renewed sense of self. Science 296:301–305

Medzhitov R (2001) Toll-like receptors and innate immunity. Nat Rev Immunol 1:135–145

Minoretti P, Gazzaruso C, Vito CD, Emanuele E, Bianchi M, Coen E, Reino M, Geroldi D (2006) Effect of the functional toll-like receptor 4 Asp299Gly polymorphism on susceptibility to late-onset Alzheimer's disease. Neurosci Lett 391:147–149

Nadeau S, Rivest S (2003) Glucocorticoids play a fundamental role in protecting the brain during innate immune response. J Neurosci 23:5536–5544

Nusslein-Volhard C, Frohnhofer HG, Lehmann R (1987) Determination of anteroposterior polarity in *Drosophila*. Science 238:1675–1681

Okamura Y, Watari M, Jerud ES, Young DW, Ishizaka ST, Rose J, Chow JC, Strauss JF 3rd (2001) The extra domain A of fibronectin activates Toll-like receptor 4. J Biol Chem 276: 10229–10233

Perry VH, Cunningham C, Holmes C (2007) Systemic infections and inflammation affect chronic neurodegeneration. Nat Rev Immunol 7:161–167

Prinz M, Garbe F, Schmidt H, Mildner A, Gutcher I, Wolter K, Piesche M, Schroers R, Weiss E, Kirschning CJ, Rochford CD, Bruck W, Becher B (2006) Innate immunity mediated by TLR9 modulates pathogenicity in an animal model of multiple sclerosis. J Clin Invest 116:456–464

Qin L, Liu Y, Wang T, Wei SJ, Block ML, Wilson B, Liu B, Hong JS (2004) NADPH oxidase mediates lipopolysaccharide-induced neurotoxicity and proinflammatory gene expression in activated microglia. J Biol Chem 279:1415–1421

Salaria S, Badkoobehi H, Rockenstein E, Crews L, Chana G, Masliah E, Everall IP (2007) Toll-like receptor pathway gene expression is associated with human immunodeficiency virus-associated neurodegeneration. J Neurovirol 13:496–503

Saura J (2007) Microglial cells in astroglial cultures: a cautionary note. J Neuroinflammation 4:26

Stefanova N, Reindl M, Neumann M, Kahle PJ, Poewe W, Wenning GK (2007) Microglial activation mediates neurodegeneration related to oligodendroglial alpha-synucleinopathy: implications for multiple system atrophy. Mov Disord 22:2196–2203

Stein D, Roth S, Vogelsang E, Nusslein-Volhard C (1991) The polarity of the dorsoventral axis in the *Drosophila* embryo is defined by an extracellular signal. Cell 65:725–735

Suzuki M, Sugimoto Y, Ohsaki Y, Ueno M, Kato S, Kitamura Y, Hosokawa H, Davies JP, Ioannou YA, Vanier MT, Ohno K, Ninomiya H (2007) Endosomal accumulation of Toll-like receptor 4 causes constitutive secretion of cytokines and activation of signal transducers and activators of transcription in Niemann–Pick disease type C (NPC) fibroblasts: a potential basis for glial cell activation in the NPC brain. J Neurosci 27:1879–1891

Trendelenburg G (2008) Acute neurodegeneration and the inflammasome: central processor for danger signals and the inflammatory response? J Cereb Blood Flow Metab 28:867–881

Ziegler G, Harhausen D, Schepers C, Hoffmann O, Rohr C, Prinz V, Konig J, Lehrach H, Nietfeld W, Trendelenburg G (2007) TLR2 has a detrimental role in mouse transient focal cerebral ischemia. Biochem Biophys Res Commun 359:574–579

Toll-Like Receptors in Spinal Cord Injury

Kristina A. Kigerl and Phillip G. Popovich

Contents

Abstract Following traumatic spinal cord injury (SCI), activated glia and inflammatory leukocytes contribute to both neurodegeneration and repair. The mechanisms that control these divergent functions are poorly understood. Toll-like receptors (TLRs) are a highly conserved family of receptors involved in pathogen recognition and host defense. However, recently it was shown that TLRs are expressed on a range of neuronal and non-neuronal cells (e.g., glia, stem/progenitor cells and leukocytes), and that nonpathogenic molecules released from sites of tissue injury, i.e., danger-associated molecular patterns (DAMPs), can activate cells via TLRs. This review will discuss how DAMPs acting at various TLRs may influence injury and repair processes of relevance to SCI, i.e., neurotoxicity, demyelination, growth cone collapse and stem/progenitor cell turnover.

K.A. Kigerl and P.G. Popovich (✉)
Center for Brain and Spinal Cord Repair, Department of Molecular Virology, Immunology, & Medical Genetics, The Ohio State University College of Medicine, Columbus, OH, USA
e-mail: Phillip.Popovich@osumc.edu

T. Kielian (ed.), *Toll-like Receptors: Roles in Infection and Neuropathology*, Current Topics in Microbiology and Immunology 336,
DOI 10.1007/978-3-642-00549-7_7,

Abbreviations

BDNF	Brain-derived neurotrophic factor
CNS	Central nervous system
CNTF	Ciliary neurotrophic factor
CSPG	Chondroitin sulfate proteoglycan
DAMPs	Danger-associated molecular patterns
DRG	Dorsal root ganglion
GGF	Glial growth factor
HMGB1	High mobility group box 1
HSP	Heat-shock protein
IL-1β	Interleukin-1 beta
IRAK	Interleukin-1 receptor-associated kinase
LPS	Lipopolysaccharide
MAG	Myelin-associated glycoprotein
MyD88	Myeloid differentiation primary response gene (88)
NF-κB	Nuclear factor kappa B
NPC	Neural progenitor cell
NT-4	Neurotrophin 4
OPC	Oligodendrocyte progenitor cell
PAMPs	Pathogen-associated molecular patterns
Poly I:C	Polyinosinic:polycytidylic acid
SCI	Spinal cord injury
SOCS	Suppressor of cytokine signaling
TLR	Toll-like receptor
TNF-α	Tumor necrosis factor alpha
Tollip	Toll-interacting protein

1 Pathology and the Inflammatory Response Triggered by Traumatic SCI

Traumatic spinal cord injury (SCI) in mammals is a devastating problem for which there is no cure. The primary injury, commonly the result of vertebral fracture or dislocation, causes contusion/compression trauma to the spinal cord. Similar types of slower compressive trauma can occur as a result of metastatic tumor invasion into the spinal canal, surgical intervention or degenerative bone disease (Prasad and Schiff 2005; Shedid and Benzel 2007; Babb and Carlson 2006). At the site of trauma, hemorrhage is grossly visible while microscopic signs of axon and myelin damage are delayed for ~24–48 h postinjury (Balentine 1978a,b; Rosenberg and Wrathall 1997). It is thought that this delayed microscopic pathology can be attenuated by blocking one or more secondary injury cascades that cause neurodegenerative changes over several days and weeks postinjury (Dusart and Schwab 1994;

Sekhon and Fehlings 2001). These secondary injury processes include ischemia, excitotoxicity and inflammation (Sekhon and Fehlings 2001).

Inflammation is a prominent and persistent feature of the injured spinal cord microenvironment. In clinical and experimental SCI, CNS macrophages, comprised of resident microglia and newly recruited blood monocytes, dominate the lesion site and nearby spinal segments within the first week postinjury (Blight 1994; Popovich et al. 1997; Popovich and Hickey 2001; Sroga et al. 2003; Kigerl et al. 2006; Chang 2007; Fleming et al. 2006). The functional significance of this response is controversial and the predominant signaling pathways involved in activating CNS macrophages remain elusive. Acute depletion or inhibition of CNS macrophages is neuroprotective after SCI (Giulian and Robertson 1990; Blight 1994; Popovich et al. 1999; Mabon et al. 2000). This is presumably because of the reduced accumulation of proinflammatory cytokines, neurotoxins and proteases released by CNS macrophages (Banati et al. 1993; Satake et al. 2000; Bao et al. 2004). Still, these same cells have proregenerative and neurotrophic potential in vitro and in vivo (Perry et al. 1987; Elkabes et al. 1996). Given the functional heterogeneity exhibited by CNS macrophages, therapeutic manipulation of this response may only be feasible once we understand the different signals that drive CNS macrophage activation in the lesion environment.

The kinetics and composition of post-traumatic SCI inflammation have been well characterized (Fleming et al. 2006; Popovich et al. 1997; Kigerl et al. 2006; Sroga et al. 2003; Chang 2007). However, little is known about the role of TLRs in shaping the functional potential of this response. Recent data suggest that TLRs expressed on immune cells and CNS cells may activate cellular and molecular cascades that are important for neuroprotection, axon regeneration and cell replacement (Kigerl et al. 2007; Schonberg et al. 2007; Cameron et al. 2007; Ma et al. 2006; Su et al. 2005; Zhang et al. 2005; Rolls et al. 2007). Here we will present a general overview of what is known about the downstream effects of TLR signaling in the context of neuron/glial cell death, survival and regeneration, with a focus on the implications of these responses in the traumatically injured spinal cord (Fig. 1).

2 TLRs Promote CNS Degeneration and Regeneration

2.1 Toll-Like Receptors and DAMPs

Toll-like receptors (TLRs) are the mammalian homologs of the Toll receptor first identified in *Drosophila* (Anderson et al. 1985). Mammalian TLRs are found on a range of cells throughout the body but have been highly scrutinized for their role in propagating innate (e.g., monocytes/macrophages; dendritic cells) and adaptive (e.g., B lymphocytes) immune cell function. In particular, TLRs recognize conserved molecular sequences found on microbes known as pathogen-associated molecular patterns (PAMPs). When PAMPs bind TLRs, NF-κB signaling is

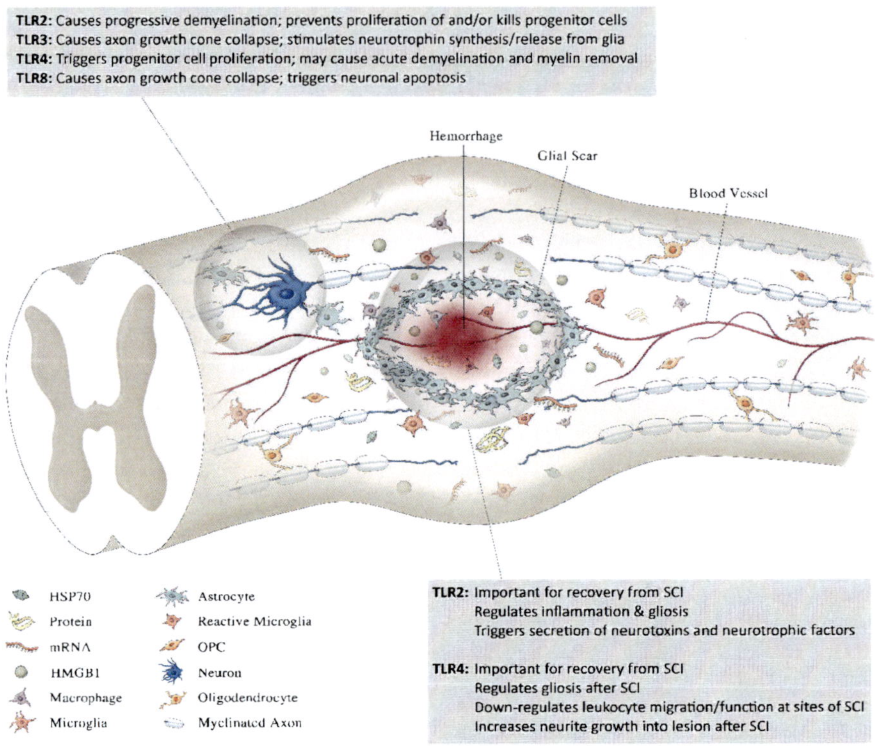

Fig. 1 A schematic illustrating known or proposed consequences of activating TLRs in the injured spinal cord. Note that the concentration of "damage-associated molecular patterns" (DAMPs) changes as a function of distance from the lesion center. This is expected to initiate divergent functions in TLR-expressing cells

initiated, resulting in the increased production of proinflammatory cytokines (Ozinsky et al. 2000; Medzhitov 2001; Nguyen et al. 2002).

In the pathological CNS, TLRs are found predominantly on microglia and macrophages (Laflamme and Rivest 2001; Zekki et al. 2002); although emerging data suggest that glia (astrocytes and oligodendrocytes), stem cells and neurons also express TLRs (Cameron et al. 2007; Ma et al. 2006; Rolls et al. 2007; Bsibsi et al. 2002; Bowman et al. 2003). It is clear that TLRs are essential for initiating host defense reactions against viral and bacterial pathogens in the CNS (Laflamme et al. 2001, 2003; Laflamme and Rivest 2001; Esen et al. 2004; Kielian et al. 2005a,b). However, TLR-mediated activation of glia, neurons and inflammatory leukocytes may also initiate or perpetuate CNS pathology in models of autoimmune demyelination, cerebral ischemia and peripheral neuropathy (Zekki et al. 2002; Maslinska et al. 2004; Tanga et al. 2005). In these latter forms of pathology, TLRs may initiate intracellular signaling by binding "danger-associated molecular patterns" (DAMPs). These *endogenous nonpathogenic* molecules that are released at sites of tissue injury can act as TLR ligands and may serve a physiological role in alerting cells to marked changes in tissue homeostasis (Zhang and Schluesener 2006; Bianchi 2007).

Many DAMPs, including heat-shock proteins (HSPs), necrotic cells, degradation products of extracellular matrix molecules (e.g., fibronectin and hyaluronic acid), high mobility group box 1 (HMGB1) and mRNA (Asea et al. 2002; Basu et al. 2000; Johnson et al. 2002; Li et al. 2001; Ohashi et al. 2000; Okamura et al. 2001; Taguchi et al. 1996; Termeer et al. 2002; Yu et al. 2006) increase at sites of SCI. For example, hyaluronan becomes overabundant in the spinal cord extracellular matrix as a consequence of astrogliosis (Struve et al. 2005). Similarly, heat shock protein 70 (HSP70) is upregulated in glia and inflammatory cells in/nearby the site of SCI within the first weeks postinjury (Mautes and Noble 2000). Other DAMPs released from necrotic cells at the site of primary trauma include HSPs, HMGB1 and mRNA (Kariko et al. 2004; Asea et al. 2002; Rovere-Querini et al. 2004). Thus, an abundance of endogenous TLR ligands exist in the injured spinal cord. As such, it is probable that these and as yet undefined TLR ligands activate a range of TLR-dependent signaling cascades in glia and immune cells, and that these signals help shape the pathophysiological sequelae of SCI. Indeed, HSP or hyaluronan can trigger TNF-α, IL-1β, and nitric oxide release from macrophages via TLR activation (Asea et al. 2002; Ohashi et al. 2000). Each of these inflammatory mediators has been implicated in post-SCI degeneration and repair (Satake et al. 2000; Nesic et al. 2001; Yune et al. 2003; Kim et al. 2001). Also, the rapid activation of NF-κB in microglia within 30 min after SCI (Bethea et al. 1998) could be triggered by DAMPs released from necrotic cells acting on TLRs (Li et al. 2001). It was recently shown that HSP60 can trigger neuronal death upon release from necrotic cells (Lehnardt et al. 2008). This effect was mediated via activation of TLR4 on microglia. Thus, early signaling mediated by select TLRs (e.g., TLR2, TLR4) could trigger a feedforward cascade of inflammatory-mediated neurodegeneration. However, a detailed review of the existing data suggests that the effects of TLR signaling are complex, and in most cases it is still premature to ascribe pro-injury or pro-repair functions to specific TLRs.

2.2 Divergent Functions are Triggered by Select TLRs

For years, mechanisms of inflammatory-mediated neurotoxicity have been modeled in vivo and in vitro by introducing LPS into the system (Bronstein et al. 1995; Kim et al. 2000; Piani and Fontana 1994). However, only recently was it shown that LPS-mediated killing of oligodendrocytes and neurons requires TLR4 signaling on microglia (Lehnardt et al. 2003, 2002). Interestingly, the toxic effects of LPS seem to be concentration dependent; higher concentrations of LPS cause microglia to adopt a neurotoxic phenotype, whereas lower concentrations create innocuous or neurite growth-promoting microglia (Li et al. 2007). These effects may have implications in traumatized spinal cord where the concentrations of TLR ligands will vary throughout the lesion—a higher concentration of DAMPs is expected at/nearby the site of injury, while lower concentrations would be found in sites removed from the lesion center (e.g., within the dorsal columns proximal to the lesion where axons are undergoing Wallerian degeneration) (Fig. 1).

Stimulation of microglia/macrophages with zymosan, a potent TLR2 ligand (Underhill et al. 1999), produces conflicting effects on mechanisms of CNS injury and repair. Specifically, zymosan-stimulated macrophages promote the growth of retinal ganglion cells in vitro and in vivo (Yin et al. 2003, 2006). This proregenerative effect was shown to be dependent on macrophages synthesizing and releasing the low molecular weight protein oncomodulin (Yin et al. 2006). However, these same macrophages also release high molecular weight proteins that are neurotoxic (Yin et al. 2003). New data from our lab show that TLR2-activated macrophages in spinal cord promote axon growth while simultaneously causing neuron and glial cell death Gensel et al. 2009. Because TLR2 forms heterodimers with either TLR1 or TLR6, ligand-dependent recruitment of distinct heterodimers may trigger unique effector functions in TLR-expressing cells (Ozinsky et al. 2000; Triantafilou et al. 2006). Alternatively, recent data indicate that, for a given cell, these different TLR2 heterodimers expand the repertoire of ligand binding but without causing divergent intracellular signaling cascades (Farhat et al. 2008). It is intriguing to speculate that DAMPs in the injured spinal cord may initiate pathogenic or reparative functions via differential recruitment of TLR2/TLR6 or TLR2/TLR1 heterodimers.

TLR3 ligands trigger the production of neurotrophic factors and anti-inflammatory cytokines from astrocytes (Bsibsi et al. 2006) but also induce growth cone collapse in growing neurons (Cameron et al. 2007). Conditioned medium from human astrocyte cultures stimulated with the synthetic TLR3 ligand, polyinosine:polycytidylic acid (poly I:C), increased neuron survival in organotypic cultures of human brain slices (Bsibsi et al. 2006). Subsequent analyses of this neurotrophic medium revealed a TLR3-dependent increase in brain-derived neurotrophic factor (BDNF), ciliary neurotrophic factor (CNTF), neurotrophin 4 (NT-4) and glial growth factors 1 & 2 (GGF1/neuregulin 1, GGF2/neuregulin). These factors are known to regulate cellular growth, differentiation, and migration of cells in the injured spinal cord (McTigue et al. 1998; Ankeny et al. 2001; Ye and Houle 1997; Kobayashi et al. 1997). TLR3-activated astrocytes also produced increased levels of interleukin 10 (Bsibsi et al. 2006), an anti-inflammatory cytokine known to provide neuroprotection and promote functional recovery after SCI (Bethea et al. 1999).

In neurons, two intracellular TLRs, TLR3 and TLR8, are concentrated in the cell body, the axon and the growth cone (Ma et al. 2006; Cameron et al. 2007; Tang et al. 2007; Lafon et al. 2006). When poly I:C or mRNA are added to DRG neurons in culture, both cause growth cone collapse, implicating TLR3 in the molecular regulation of axonal plasticity and regeneration (Cameron et al. 2007). Moreover, when poly I:C is injected into neonatal mice, sensory axon growth is inhibited and these mice develop sensorimotor deficits. The specificity of these effects was proven when poly I:C failed to elicit anatomical or functional deficits in developing

TLR3 knockout mice (Cameron et al. 2007). Given the large amount of cell necrosis that occurs at the site of a SCI, it is likely that axons that are attempting to regenerate will encounter mRNA. Because mRNA is an endogenous ligand for TLR3 (Kariko et al. 2004), it must be considered alongside myelin proteins (e.g., NOGO, MAG) and CSPGs as another component of the molecular barrier to axon regeneration in the injured spinal cord (Silver and Miller 2004). Given that astrocytes also become activated via TLR3 (see above), it will be interesting to see if their ability to secrete neurotrophins, a presumably beneficial aspect of CNS repair, is accompanied by a concomitant increase in the expression of molecules that are potent inhibitors of axon growth (e.g., CSPGs).

Similar to the effects of TLR3 on DRG neurons, activation of TLR8 in cortical neurons inhibits neurite outgrowth (Ma et al. 2006). However, TLR8 ligation simultaneously triggers apoptosis—an effect that was not observed in DRG neurons after TLR3 stimulation (Ma et al. 2006). This implies that TLRs can exert diverse functions in subsets of the same cell type in response to similar TLR signals. In this context, it would be interesting to know if microglia and monocyte-derived macrophages mount distinct responses to endogenous (e.g., HMGB1) or pathogenic (e.g., LPS) TLR ligands.

3 The Effects of Activating TLRs In Vivo in Spinal Cord

In early studies from our group, simplified models were used to study the consequences of intraspinal macrophage activation. Importantly, this model allowed TLR-mediated CNS macrophage activation without the confounding effects of trauma or hemorrhage. When microinjected into the intact brain or spinal cord, the TLR2 agonist zymosan triggers a florid CNS macrophage response within three days (Fitch et al. 1999; Popovich et al. 2002). Axonal degeneration and demyelination co-localize with the CNS macrophage response, creating lesions that are reminiscent of those caused by blunt trauma (Fitch et al. 1999). Importantly, just as we have described in models of traumatic SCI, the intraspinal pathology caused by zymosan microinjection is significantly reduced if circulating monocytes are depleted during the first week postinjection (unpublished data). Collectively, these data suggest that CNS macrophages release neurotoxic mediators when activated by TLR2 (see above). More recently, Schonberg et al. used an identical protocol to compare the effects of intraspinal TLR2 and TLR4 activation on oligodendrocyte progenitor cell (OPC) survival and oligodendrocyte formation (Schonberg et al. 2007). Both zymosan (TLR2) and LPS (TLR4) elicit robust CNS macrophage activation; however, only zymosan caused significant axonal and myelin damage with evidence of collateral/bystander demyelination at and near the injection site. In contrast, TLR4-activated CNS macrophages co-localized to regions of marked OPC proliferation and new oligodendrocyte formation (Schonberg et al. 2007). Increased oligogenesis was attributed to enhanced CNTF production by glia at the site of injection.

Not all studies have reported salutary effects when LPS is injected into the spinal cord. Felts et al. showed that LPS injections into the spinal cord dorsal column of naïve rats elicits inflammation culminating within seven days of injection with the formation of large primary demyelinating lesions (Felts et al. 2005). Within 28 days these lesions become remyelinated by Schwann cells. Differences in the volume of LPS that was injected may partially explain why these apparently similar models of intraspinal inflammation produced distinct results. Although both studies injected identical nanogram quantities of LPS, Schonberg et al. injected a volume that was 1/10 that reported by Felts et al. (200 nl compared to 2 μl). It is reasonable to speculate that the larger injection volume creates an artificial edema that primes glial activation and/or alters intercellular communication. Consequently, the CNS macrophage response that is elicited by LPS may be fundamentally different from that created when smaller volumes are injected. Overall, these data suggest that the consequence of TLR signaling in CNS macrophages will be a function of the ligand *and* the environment in which the TLR-expressing cells are activated.

When delivered systemically, TLR ligands can induce proliferation of neural progenitor cells (NPCs). Su et al. showed proliferation of nestin-positive NPCs in the spinal cord after intraperitoneal injection of LPS, R-848 (TLR7/8 ligand) or poly I:C (TLR3). CpG-ODN (TLR9 ligand) injections did not trigger cell proliferation. Following injections of R-848 or poly I:C, microglia undergo phenotypic and morphological transformation with some evidence of proliferation; however, most responding cells had an NPC phenotype (Su et al. 2005). Recently, adult hippocampal NPCs were shown to express TLR2 and TLR4. It was suggested that these TLRs have opposing roles in regulating NPC proliferation and differentiation; TLR2 deficiency impairs neurogenesis while TLR4 deficiency augments proliferation and neurogenesis (Rolls et al. 2007). Given the prominent role of Toll in *Drosophila* development, it is logical to predict a role for TLRs in regulating proliferation and differentiation of CNS progenitor cells. If we can learn how TLRs control neural and glial progenitor cell fate, it may be possible to manipulate these receptors as part of a cell replacement therapy for SCI or other neurological diseases (Thuret et al. 2006; Bradbury and McMahon 2006).

4 TLR Activation and SCI

Few studies have directly examined the consequences of TLR activation in the traumatically injured spinal cord. In a model of spinal contusion injury we used gene expression profiling to document rapid and sustained induction of TLRs 1, 2, 4, 6, 7 and 9. Most TLRs were expressed greater than twofold above sham-injury controls as early as one day postinjury, with changes lasting for up to four weeks. In addition, several TLR-associated signaling molecules (i.e., MyD88, IRAK-4) were increased during the first 1–2 weeks postinjury (Kigerl et al. 2007). These broad and prolonged changes in the expression of TLRs and related signaling molecules suggest a prominent role for this receptor family in regulating post-traumatic inflammation, regeneration and cell death/survival. Importantly, in this model, TLRs are induced in the absence of exogenous pathogens. What remains unclear is the range of cells that upregulate TLRs after SCI, and which DAMPs participate in TLR-dependent signaling.

From injured spinal cord sections labeled for CNS macrophages or astrocytes, we used laser capture microdissection to examine cell-specific TLR4 and TLR2 mRNA expression. We found that CNS macrophages are the predominant TLR4-expressing cells; however, both CNS macrophages and astrocytes expressed TLR2 (Kigerl et al. 2007).

To reveal the functional significance of SCI-mediated TLR induction, we evaluated neuropathological changes and functional recovery in SCI mice deficient in TLR4 or TLR2 signaling. We limited our focus to these TLRs since most endogenous ligands activate innate immune cells via TLR2 or TLR4 (Ohashi et al. 2000; Li et al. 2001; Okamura et al. 2001; Termeer et al. 2002; Asea et al. 2002). Because TLR-activated macrophages initiate and amplify inflammatory responses, we predicted that pathology would be attenuated and neurological function improved in mice deficient in TLR2 or TLR4 signaling. Indeed, despite the conflicting effects of TLR2- and TLR4-mediated CNS macrophage activation on axon pathology, neuron survival and demyelination (Fitch et al. 1999; Lehnardt et al. 2002; Felts et al. 2005), it is generally accepted that acute suppression or depletion of macrophages is neuroprotective after SCI (Popovich et al. 1999; Blight 1994; Giulian and Robertson 1990; Gris et al. 2004). In opposition to our hypothesis, our data showed that deficiencies in TLR2 and TLR4 signaling exacerbated post-traumatic accumulation of CNS macrophages. This was accompanied by sustained functional impairment, aberrant patterns of gliosis and excessive axon/myelin pathology (Kigerl et al. 2007). However, consistent with the role of TLR2 and TLR4 in triggering proinflammatory cytokine synthesis, we found that expression of IL-1β and TNFα mRNA was attenuated in mice deficient in TLR2 and TLR4 signaling. Despite this, CNS macrophages continued to accumulate in and near the site of injury for up to six weeks (latest time examined) in TLR-deficient mice (Kigerl et al. 2007). From these data, we conclude that instead of priming activation signals in CNS macrophages, DAMPs may play an important role in triggering immune-regulatory cascades and/or initiating neuroprotective inflammatory functions (see below).

In the mid-1950s, an unexpected discovery revealed a potential neuroprotective role for TLR4 activation in the injured spinal cord. While attempting to study neural centers of temperature regulation, Windle and colleagues discovered that administration of a pyrogenic bacterial polysaccharide restored sensation in spinalized dogs (Windle and Chambers 1950). It was later shown that this pyrogen preparation induced marked intraspinal inflammation with a concomitant decrease in fibrosis and astrogliosis (Clemente and Windle 1954). Since then, several groups have administered LPS as a therapeutic agent in models of SCI (Guth et al. 1994a, b; Davis et al. 2005; Vallieres et al. 2006). Guth et al. showed that LPS reduced lesion cavitation, but surprisingly also reduced macrophage accumulation at the injury site (Guth et al. 1994a). These changes were associated with an increase in axon density in the lesion site. What remains unclear in this model is whether LPS influenced macrophages in the periphery, in the CNS, or both. Also, were the effects of LPS observed by Guth et al. specific for CNS macrophages? Indeed, as we have stated above, LPS (or endotoxin) could have widespread effects on multiple cell types and these effects may be dose dependent. More recent studies support this point. Davis et al. showed that when injected systemically at increasingly higher concentrations, LPS decreases the number of circulating monocytes and neutrophils, resulting in fewer neutrophils and monocytes infiltrating the injured spinal cord (Davis et al. 2005). Systemic LPS may also limit the permeability of the injured blood–spinal cord barrier (Davis et al. 2005).

These data, when interpreted together with our data showing increased CNS macrophage responses in TLR4-deficient mice after SCI (Kigerl et al. 2007), support the hypothesis that TLR4 activates then downregulates CNS inflammation. This could occur by triggering "autoregulatory apoptosis." Indeed, TLR4 activation of microglia is known to trigger the release of proinflammatory cytokines, but to also cause apoptosis (Jung et al. 2005). TLR4 activation also signals the activation of several inhibitory molecules involved in suppressing TLR and/or cytokine signaling. These proteins include suppressor of cytokine signaling 1 (SOCS1), which inhibits TLR and cytokine signaling, and several Toll/IL-1-specific inhibitors, including IRAK-M and Tollip (Kobayashi et al. 2002; Kinjyo et al. 2002; Nakagawa et al. 2002; Zhang and Ghosh 2002; Liew et al. 2005). It is likely that these negative TLR4 regulators have evolved to prevent tissue damage caused by excessive inflammation. In fact, IRAK-M expression has been linked to the sepsis-induced immunosuppression that occurs in patients recovering from septic shock (Deng et al. 2006; Reddy et al. 2001). A similar mechanism may explain why we have observed increased inflammation and tissue damage in TLR4-deficient mice after SCI.

5 Conclusions

It is clear that TLRs do far more than recognize and initiate host-defense reactions against pathogens. TLRs regulate various cellular processes that influence the balance between tissue damage and repair in the traumatically injured spinal cord—a site where pathogens are rare or do not exist. The massive induction of multiple TLRs in these sterile SCI lesions emphasizes this point and raises another: even though TLR receptors and their associated signaling pathways are often studied in isolation, it is unlikely that a single TLR will be engaged in a single cell type at or nearby the site of SCI. Even if this were the case, unique effector functions would be expected to be stimulated by a specific TLR as a result of varying types and concentrations of DAMPs. The apparent redundancy and breadth of TLR signaling may also explain why neuroprotective and neurotoxic effects have been described for specific TLRs (e.g., TLR4). We are at the early stages of understanding these receptors and how they may impact post-SCI neurodegeneration and repair. Still, the emerging data showing that they help coordinate post-traumatic inflammation, gliosis, neuron survival, axon regeneration and stem/progenitor cell survival/differentiation argues that they should receive increased attention as therapeutic targets for treating SCI and related neurological disorders.

References

Anderson KV, Bokla L, Nusslein-Volhard C (1985) Establishment of dorsal-ventral polarity in the *Drosophila* embryo: the induction of polarity by the Toll gene product. Cell 42:791–798

Ankeny DP, McTigue DM, Guan Z, Yan Q, Kinstler O, Stokes BT, Jakeman LB (2001) Pegylated brain-derived neurotrophic factor shows improved distribution into the spinal cord and stimulates locomotor activity and morphological changes after injury. Exp Neurol 170:85–100

Asea A, Rehli M, Kabingu E, Boch JA, Bare O, Auron PE, Stevenson MA, Calderwood SK (2002) Novel signal transduction pathway utilized by extracellular HSP70: role of toll-like receptor (TLR) 2 and TLR4. J Biol Chem 277:15028–15034

Babb A, Carlson WO (2006) Spinal stenosis. S D Med 59:103–105

Balentine JD (1978a) Pathology of experimental spinal cord trauma. I. The necrotic lesion as a function of vascular injury. Lab Invest 39:236–253

Balentine JD (1978b) Pathology of experimental spinal cord trauma. II Ultrastructure of axons and myelin. Lab Invest 39:254–266

Banati RB, Gehrmann J, Schubert P, Kreutzberg GW (1993) Cytotoxicity of microglia. Glia 7:111–118

Bao F, Chen Y, Dekaban GA, Weaver LC (2004) Early anti-inflammatory treatment reduces lipid peroxidation and protein nitration after spinal cord injury in rats. J Neurochem 88:1335–1344

Basu S, Binder RJ, Suto R, Anderson KM, Srivastava PK (2000) Necrotic but not apoptotic cell death releases heat shock proteins, which deliver a partial maturation signal to dendritic cells and activate the NF-kappa B pathway. Int Immunol 12:1539–1546

Bethea JR, Castro M, Keane RW, Lee TT, Dietrich WD, Yezierski RP (1998) Traumatic spinal cord injury induces nuclear factor-kB activation. J Neurosci 18:3251–3260

Bethea JR, Nagashima H, Acosta MC, Briceno C, Gomez F, Marcillo AE, Loor K, Green J, Dietrich WD (1999) Systemically administered interleukin-10 reduces tumor necrosis factor-alpha production and significantly improves functional recovery following traumatic spinal cord injury in rats. J Neurotrauma 16:851–863

Bianchi ME (2007) DAMPs, PAMPs and alarmins: all we need to know about danger. J Leukoc Biol 81:1–5

Blight AR (1994) Effects of silica on the outcome from experimental spinal cord injury: implication of macrophages in secondary tissue damage. Neuroscience 60:263–273

Bowman CC, Rasley A, Tranguch SL, Marriott I (2003) Cultured astrocytes express toll-like receptors for bacterial products. Glia 43:281–291

Bradbury EJ, McMahon SB (2006) Spinal cord repair strategies: why do they work? Nat Rev Neurosci 7:644–653

Bronstein DM, Perez-Otano I, Sun V, Mullis Sawin SB, Chan J, Wu GC, Hudson PM, Kong LY, Hong JS, McMillian MK (1995) Glia-dependent neurotoxicity and neuroprotection in mesencephalic cultures. Brain Res 704:112–116

Bsibsi M, Persoon-Deen C, Verwer RW, Meeuwsen S, Ravid R, van Noort JM (2006) Toll-like receptor 3 on adult human astrocytes triggers production of neuroprotective mediators. Glia 53:688–695

Bsibsi M, Ravid R, Gveric D, van Noort JM (2002) Broad expression of Toll-like receptors in the human central nervous system. J Neuropathol Exp Neurol 61:1013–1021

Cameron JS, Alexopoulou L, Sloane JA, DiBernardo AB, Ma Y, Kosaras B, Flavell R, Strittmatter SM, Volpe J, Sidman R, Vartanian T (2007) Toll-like receptor 3 is a potent negative regulator of axonal growth in mammals. J Neurosci 27:13033–13041

Chang HT (2007) Subacute human spinal cord contusion: few lymphocytes and many macrophages. Spinal Cord 45:174–182

Clemente CD, Windle WF (1954) Regeneration of severed nerve fibers in the spinal cord of the adult cat. J Comp Neurol 101:691–731

Davis AE, Campbell SJ, Wilainam P, Anthony DC (2005) Post-conditioning with lipopolysaccharide reduces the inflammatory infiltrate to the injured brain and spinal cord: a potential neuroprotective treatment. Eur J Neurosci 22:2441–2450

Deng JC, Cheng G, Newstead MW, Zeng X, Kobayashi K, Flavell RA, Standiford TJ (2006) Sepsis-induced suppression of lung innate immunity is mediated by IRAK-M. J.Clin.Invest 116:2532–2542

Dusart I, Schwab ME (1994) Secondary cell death and the inflammatory reaction after dorsal hemisection of the rat spinal cord. Eur J Neurosci 6:712–724

Elkabes S, DiCicco-Bloom EM, Black IB (1996) Brain microglia/macrophages express neurotrophins that selectively regulate microglial proliferation and function. J Neurosci 16:2508–2521

Esen N, Tanga FY, DeLeo JA, Kielian T (2004) Toll-like receptor 2 (TLR2) mediates astrocyte activation in response to the Gram-positive bacterium *Staphylococcus aureus*. J Neurochem 88:746–758

Farhat K, Riekenberg S, Heine H, Debarry J, Lang R, Mages J, Buwitt-Beckmann U, Roschmann K, Jung G, Wiesmuller KH, Ulmer AJ (2008) Heterodimerization of TLR2 with TLR1 or TLR6 expands the ligand spectrum but does not lead to differential signaling. J Leukoc Biol 83:692–701

Felts PA, Woolston AM, Fernando HB, Asquith S, Gregson NA, Mizzi OJ, Smith KJ (2005) Inflammation and primary demyelination induced by the intraspinal injection of lipopolysaccharide. Brain 128:1649–1666

Fitch MT, Doller C, Combs CK, Landreth GE, Silver J (1999) Cellular and molecular mechanisms of glial scarring and progressive cavitation: in vivo and in vitro analysis of inflammation-induced secondary injury after CNS trauma. J Neurosci 19:8182–8198

Fleming JC, Norenberg MD, Ramsay DA, Dekaban GA, Marcillo AE, Saenz AD, Pasquale-Styles M, Dietrich WD, Weaver LC (2006) The cellular inflammatory response in human spinal cords after injury. Brain 129:3249–3269

Gensel SC, Nakamura S, Van Rooijen N, Ankeny DP, Popovich PG (2009) Macrophages promote axon regeneration with concurrent neurotoxicity. J Neurosci 29(12):3956–68

Giulian D, Robertson C (1990) Inhibition of mononuclear phagocytes reduces ischemic injury in the spinal cord. Ann Neurol 27:33–42

Gris D, Marsh DR, Oatway MA, Chen Y, Hamilton EF, Dekaban GA, Weaver LC (2004) Transient blockade of the CD11d/CD18 integrin reduces secondary damage after spinal cord injury, improving sensory, autonomic, and motor function. J Neurosci 24:4043–4051

Guth L, Zhang Z, DiProspero NA, Joubin K, Fitch MT (1994a) Spinal cord injury in the rat: treatment with bacterial lipopolysaccharide and indomethacin enhances cellular repair and locomotor function. Exp Neurol 126:76–87

Guth L, Zhang Z, Roberts E (1994b) Key role for pregnenolone in combination therapy that promotes recovery after spinal cord injury. Proc Natl Acad Sci USA 91:12308–12312

Johnson GB, Brunn GJ, Kodaira Y, Platt JL (2002) Receptor-mediated monitoring of tissue well-being via detection of soluble heparan sulfate by Toll-like receptor 4. J Immunol 168: 5233–5239

Jung DY, Lee H, Jung BY, Ock J, Lee MS, Lee WH, Suk K (2005) TLR4, but not TLR2, signals autoregulatory apoptosis of cultured microglia: a critical role of IFN-beta as a decision maker. J Immunol 174:6467–6476

Kariko K, Ni H, Capodici J, Lamphier M, Weissman D (2004) mRNA is an endogenous ligand for Toll-like receptor 3. J Biol Chem 279:12542–12550

Kielian T, Esen N, Bearden ED (2005a) Toll-like receptor 2 (TLR2) is pivotal for recognition of *S. aureus* peptidoglycan but not intact bacteria by microglia. Glia 49:567–576

Kielian T, Haney A, Mayes PM, Garg S, Esen N (2005b) Toll-like receptor 2 modulates the proinflammatory milieu in *Staphylococcus aureus*-induced brain abscess. Infect Immun 73:7428–7435

Kigerl KA, Lai W, Rivest S, Hart RP, Satoskar AR, Popovich PG (2007) Toll-like receptor (TLR)-2 and TLR-4 regulate inflammation, gliosis, and myelin sparing after spinal cord injury. J Neurochem 102:37–50

Kigerl KA, McGaughy VM, Popovich PG (2006) Comparative analysis of lesion development and intraspinal inflammation in four strains of mice following spinal contusion injury. Comp Neurol 494:578–594

Kim GM, Xu J, Xu J, Song SK, Yan P, Ku G, Xu XM, Hsu CY (2001) Tumor necrosis factor receptor deletion reduces nuclear factor-kappaB activation, cellular inhibitor of apoptosis protein 2 expression, and functional recovery after traumatic spinal cord injury. J Neurosci 21:6617–6625

Kim WG, Mohney RP, Wilson B, Jeohn GH, Liu B, Hong JS (2000) Regional difference in susceptibility to lipopolysaccharide-induced neurotoxicity in the rat brain: role of microglia. J Neurosci 20:6309–6316

Kinjyo I, Hanada T, Inagaki-Ohara K, Mori H, Aki D, Ohishi M, Yoshida H, Kubo M, Yoshimura A (2002) SOCS1/JAB is a negative regulator of LPS-induced macrophage activation. Immunity 17:583–591

Kobayashi K, Hernandez LD, Galan JE, Janeway CA Jr, Medzhitov R, Flavell RA (2002) IRAK-M is a negative regulator of Toll-like receptor signaling. Cell 110:191–202

Kobayashi NR, Fan DP, Giehl KM, Bedard AM, Wiegand SJ, Tetzlaff W (1997) BDNF and NT-4/5 prevent atrophy of rat rubrospinal neurons after cervical axotomy, stimulate GAP-43 and Talpha1-tubulin mRNA expression, and promote axonal regeneration. J Neurosci 17:9583–9595

Laflamme N, Echchannaoui H, Landmann R, Rivest S (2003) Cooperation between toll-like receptor 2 and 4 in the brain of mice challenged with cell wall components derived from gram-negative and gram-positive bacteria. Eur J Immunol 33:1127–1138

Laflamme N, Rivest S (2001) Toll-like receptor 4: the missing link of the cerebral innate immune response triggered by circulating gram-negative bacterial cell wall components. FASEB J 15:155–163

Laflamme N, Soucy G, Rivest S (2001) Circulating cell wall components derived from gram-negative, not gram- positive, bacteria cause a profound induction of the gene-encoding Toll-like receptor 2 in the CNS. J Neurochem 79:648–657

Lafon M, Megret F, Lafage M, Prehaud C (2006) The innate immune facet of brain: human neurons express TLR-3 and sense viral dsRNA. J Mol Neurosci 29:185–194

Lehnardt S, Lachance C, Patrizi S, Lefebvre S, Follett PL, Jensen FE, Rosenberg PA, Volpe JJ, Vartanian T (2002) The toll-like receptor TLR4 is necessary for lipopolysaccharide-induced oligodendrocyte injury in the CNS. J Neurosci 22:2478–2486

Lehnardt S, Massillon L, Follett P, Jensen FE, Ratan R, Rosenberg PA, Volpe JJ, Vartanian T (2003) Activation of innate immunity in the CNS triggers neurodegeneration through a Toll-like receptor 4-dependent pathway. Proc Natl Acad Sci USA 100:8514–8519

Lehnardt S, Schott E, Trimbuch T, Laubisch D, Krueger C, Wulczyn G, Nitsch R, Weber JR (2008) A vicious cycle involving release of heat shock protein 60 from injured cells and activation of toll-like receptor 4 mediates neurodegeneration in the CNS. J Neurosci 28:2320–2331

Li L, Lu J, Tay SS, Moochhala SM, He BP (2007) The function of microglia, either neuroprotection or neurotoxicity, is determined by the equilibrium among factors released from activated microglia in vitro. Brain Res 1159:8–17

Li M, Carpio DF, Zheng Y, Bruzzo P, Singh V, Ouaaz F, Medzhitov RM, Beg AA (2001) An essential role of the NF-kappa B/Toll-like receptor pathway in induction of inflammatory and tissue-repair gene expression by necrotic cells. J Immunol 166:7128–7135

Liew FY, Xu D, Brint EK, O'Neill LA (2005) Negative regulation of toll-like receptor-mediated immune responses. Nat Rev Immunol 5:446–458

Ma Y, Li J, Chiu I, Wang Y, Sloane JA, Lu J, Kosaras B, Sidman RL, Volpe JJ, Vartanian T (2006) Toll-like receptor 8 functions as a negative regulator of neurite outgrowth and inducer of neuronal apoptosis. J Cell Biol 175:209–215

Mabon PJ, Weaver LC, Dekaban GA (2000) Inhibition of monocyte/macrophage migration to a spinal cord injury site by an antibody to the integrin alphaD: a potential new anti-inflammatory treatment. Exp Neurol 166:52–64

Maslinska D, Laure-Kamionowska M, Maslinski S (2004) Toll-like receptors in rat brains injured by hypoxic-ischaemia or exposed to staphylococcal alpha-toxin. Folia Neuropathol 42:125–132

Mautes AE, Noble LJ (2000) Co-induction of HSP70 and heme oxygenase-1 in macrophages and glia after spinal cord contusion in the rat. Brain Res 883(2):233–237
McTigue DM, Horner PJ, Stokes BT, Gage FH (1998) Neurotrophin-3 and brain-derived neurotrophic factor induce oligodendrocyte proliferation and myelination of regenerating axons in the contused adult rat spinal cord. J Neurosci 18:5354–5365
Medzhitov R (2001) Toll-like receptors and innate immunity. Nat Rev Immunol 1:135–145
Nakagawa R, Naka T, Tsutsui H, Fujimoto M, Kimura A, Abe T, Seki E, Sato S, Takeuchi O, Takeda K, Akira S, Yamanishi K, Kawase I, Nakanishi K, Kishimoto T (2002) SOCS-1 participates in negative regulation of LPS responses. Immunity 17:677–687
Nesic O, Xu GY, McAdoo D, High KW, Hulsebosch C, Perez-Pol R (2001) IL-1 receptor antagonist prevents apoptosis and caspase-3 activation after spinal cord injury. J Neurotrauma 18:947–956
Nguyen MD, Julien JP, Rivest S (2002) Innate immunity: the missing link in neuroprotection and neurodegeneration? Nat Rev Neurosci 3:216–227
Ohashi K, Burkart V, Flohe S, Kolb H (2000) Cutting edge: heat shock protein 60 is a putative endogenous ligand of the toll-like receptor-4 complex. J Immunol 164:558–561
Okamura Y, Watari M, Jerud ES, Young DW, Ishizaka ST, Rose J, Chow JC, Strauss JF 3rd (2001) The extra domain A of fibronectin activates Toll-like receptor 4. J Biol Chem 276:10229–10233
Ozinsky A, Underhill DM, Fontenot JD, Hajjar AM, Smith KD, Wilson CB, Schroeder L, Aderem A (2000) The repertoire for pattern recognition of pathogens by the innate immune system is defined by cooperation between toll-like receptors. Proc Natl Acad Sci USA 97:13766–13771
Perry VH, Brown MC, Gordon S (1987) The macrophage response to central and peripheral nerve injury. A possible role for macrophages in regeneration. J Exp Med 165:1218–1223
Piani D, Fontana A (1994) Involvement of the cystine transport system xc- in the macrophage-induced glutamate-dependent cytotoxicity to neurons. J Immunol 152:3578–3585
Popovich PG, Guan Z, McGaughy V, Fisher L, Hickey WF, Basso DM (2002) The neuropathological and behavioral consequences of intraspinal microglial/macrophage activation. J Neuropathol Exp Neurol 61:623–633
Popovich PG, Guan Z, Wei P, Huitinga I, van Rooijen N, Stokes BT (1999) Depletion of hematogenous macrophages promotes partial hindlimb recovery and neuroanatomical repair after experimental spinal cord injury. Exp Neurol 158:351–365
Popovich PG, Hickey WF (2001) Bone marrow chimeric rats reveal the unique distribution of resident and recruited macrophages in the contused rat spinal cord. J Neuropathol Exp Neurol 60:676–685
Popovich PG, Wei P, Stokes BT (1997) The cellular inflammatory response after spinal cord injury in Sprague–Dawley and Lewis rats. J Comp Neurol 377:443–464
Prasad D, Schiff D (2005) Malignant spinal-cord compression. Lancet Oncol 6:15–24
Reddy RC, Chen GH, Tekchandani PK, Standiford TJ (2001) Sepsis-induced immunosuppression: from bad to worse. Immunol Res 24:273–287
Rolls A, Shechter R, London A, Ziv Y, Ronen A, Levy R, Schwartz M (2007) Toll-like receptors modulate adult hippocampal neurogenesis. Nat Cell Biol 9:1081–1088
Rosenberg LJ, Wrathall JR (1997) Quantitative analysis of acute axonal pathology in experimental spinal cord contusion. J Neurotrauma 14:823–838
Rovere-Querini P, Capobianco A, Scaffidi P, Valentinis B, Catalanotti F, Giazzon M, Dumitriu IE, Muller S, Iannacone M, Traversari C, Bianchi ME, Manfredi AA (2004) HMGB1 is an endogenous immune adjuvant released by necrotic cells. EMBO Rep 5:825–830
Satake K, Matsuyama Y, Kamiya M, Kawakami H, Iwata H, Adachi K, Kiuchi K (2000) Nitric oxide via macrophage iNOS induces apoptosis following traumatic spinal cord injury. Brain Res Mol Brain Res 85:114–122
Schonberg DL, Popovich PG, McTigue DM (2007) Oligodendrocyte generation is differentially influenced by Toll-like receptor (TLR) 2 and TLR4-mediated intraspinal macrophage activation. J Neuropathol Exp Neurol 66:1124–1135

Sekhon LH, Fehlings MG (2001) Epidemiology, demographics, and pathophysiology of acute spinal cord injury. Spine 26:S2–S12

Shedid D, Benzel EC (2007) Cervical spondylosis anatomy: pathophysiology and biomechanics. Neurosurgery 60:S7–S13

Silver J, Miller JH (2004) Regeneration beyond the glial scar. Nat Rev Neurosci 5:146–156

Sroga JM, Jones TB, Kigerl KA, McGaughy VM, Popovich PG (2003) Rats and mice exhibit distinct inflammatory reactions after spinal cord injury. J Comp Neurol 462:223–240

Struve J, Maher PC, Li YQ, Kinney S, Fehlings MG, Kuntz C, Sherman LS (2005) Disruption of the hyaluronan-based extracellular matrix in spinal cord promotes astrocyte proliferation. Glia 52:16–24

Su Y, Zhang Z, Trautmann K, Xu S, Schluesener HJ (2005) TLR and NOD2 ligands induce cell proliferation in the rat intact spinal cord. J Neuropathol Exp Neurol 64:991–997

Taguchi T, Mitcham JL, Dower SK, Sims JE, Testa JR (1996) Chromosomal localization of TIL, a gene encoding a protein related to the *Drosophila* transmembrane receptor Toll, to human chromosome 4p14. Genomics 32:486–488

Tang SC, Arumugam TV, Xu X, Cheng A, Mughal MR, Jo DG, Lathia JD, Siler DA, Chigurupati S, Ouyang X, Magnus T, Camandola S, Mattson MP (2007) Pivotal role for neuronal Toll-like receptors in ischemic brain injury and functional deficits. Proc Natl Acad Sci USA 104:13798–13803

Tanga FY, Nutile-McMenemy N, DeLeo JA (2005) The CNS role of Toll-like receptor 4 in innate neuroimmunity and painful neuropathy. Proc Natl Acad Sci USA 102:5856–5861

Termeer C, Benedix F, Sleeman J, Fieber C, Voith U, Ahrens T, Miyake K, Freudenberg M, Galanos C, Simon JC (2002) Oligosaccharides of hyaluronan activate dendritic cells via toll-like receptor 4. J Exp Med 195:99–111

Thuret S, Moon LD, Gage FH (2006) Therapeutic interventions after spinal cord injury. Nat Rev Neurosci 7:628–643

Triantafilou M, Gamper FG, Haston RM, Mouratis MA, Morath S, Hartung T, Triantafilou K (2006) Membrane sorting of toll-like receptor (TLR)-2/6 and TLR2/1 heterodimers at the cell surface determines heterotypic associations with CD36 and intracellular targeting. J Biol Chem 281:31002–31011

Underhill DM, Ozinsky A, Hajjar AM, Stevens A, Wilson CB, Bassetti M, Aderem A (1999) The Toll-like receptor 2 is recruited to macrophage phagosomes and discriminates between pathogens. Nature 401:811–815

Vallieres N, Berard JL, David S, Lacroix S (2006) Systemic injections of lipopolysaccharide accelerates myelin phagocytosis during Wallerian degeneration in the injured mouse spinal cord. Glia 53:103–113

Windle WF, Chambers WW (1950) Regeneration in the spinal cord of the cat and dog. J Comp Neurol 93:241–257

Ye JH, Houle JD (1997) Treatment of the chronically injured spinal cord with neurotrophic factors can promote axonal regeneration from supraspinal neurons. Exp Neurol 143:70–81

Yin Y, Cui Q, Li Y, Irwin N, Fischer D, Harvey AR, Benowitz LI (2003) Macrophage-derived factors stimulate optic nerve regeneration. J Neurosci 23:2284–2293

Yin Y, Henzl MT, Lorber B, Nakazawa T, Thomas TT, Jiang F, Langer R, Benowitz LI (2006) Oncomodulin is a macrophage-derived signal for axon regeneration in retinal ganglion cells. Nat Neurosci 9:843–852

Yu M, Wang H, Ding A, Golenbock DT, Latz E, Czura CJ, Fenton MJ, Tracey KJ, Yang H (2006) HMGB1 signals through toll-like receptor (TLR) 4 and TLR2. Shock 26:174–179

Yune TY, Chang MJ, Kim SJ, Lee YB, Shin SW, Rhim H, Kim YC, Shin ML, Oh YJ, Han CT, Markelonis GJ, Oh TH (2003) Increased production of tumor necrosis factor-alpha induces apoptosis after traumatic spinal cord injury in rats. J Neurotrauma 20:207–219

Zekki H, Feinstein DL, Rivest S (2002) The clinical course of experimental autoimmune encephalomyelitis is associated with a profound and sustained transcriptional activation of the genes encoding toll-like receptor 2 and CD14 in the mouse CNS. Brain Pathol 12:308–319

Zhang G, Ghosh S (2002) Negative regulation of toll-like receptor-mediated signaling by Tollip. J Biol Chem 277:7059–7065

Zhang Z, Schluesener HJ (2006) Mammalian toll-like receptors: from endogenous ligands to tissue regeneration. Cell Mol Life Sci 63:2901–2907

Zhang Z, Trautmann K, Schluesener HJ (2005) Microglia activation in rat spinal cord by systemic injection of TLR3 and TLR7/8 agonists. J Neuroimmunol 164:154–160

Toll-Like Receptors in Alzheimer's Disease

Gary E. Landreth and Erin G. Reed-Geaghan

Contents

Abstract Alzheimer's disease (AD) is characterized by the formation of insoluble deposits of β-amyloid (Aβ) within the parenchyma of the brain. These deposits are associated with a robust microglia-mediated inflammatory response. Recent work has demonstrated that Toll-like receptors (TLRs) participate in this inflammatory response. This chapter reviews the mechanisms whereby TLRs contribute to the induction of a microglial inflammatory response to promote AD pathogenesis. Specifically, the involvement of CD14 and the TLRs in microglial activation is delineated. The TLR-mediated microglial response has beneficial roles in stimulating phagocytosis as well as detrimental roles in the Aβ-stimulated release of neurotoxic products.

Abbreviations

Aβ	β-Amyloid
AD	Alzheimer's disease
APP	Amyloid precursor protein

G.E. Landreth (✉) and E.G. Reed-Geaghan
Alzheimer Research Laboratory, Department of Neurosciences,
Case Western Reserve University, Cleveland, OH, 44106, USA
e-mail: gel2@case.edu

T. Kielian (ed.), *Toll-like Receptors: Roles in Infection and Neuropathology*,
Current Topics in Microbiology and Immunology 336,
DOI 10.1007/978-3-642-00549-7_8,

BACE	β-Secretase
FPRL1	Formyl peptide receptor-like 1
IFN	Interferon
IL	Interleukin
IRAK	IL-1 receptor-associated kinase
IRF	Interferon regulatory factor
MAPK	Mitogen-activated protein kinase
MHC II	Major histocompatibility complex II
mFPR2	Murine formyl peptide receptor 2
MyD88	Myeloid differentiation primary response gene 88
NSAIDs	Nonsteroidal anti-inflammatory drugs
PRRs	Pattern recognition receptors
PAMPs	Pathogen-associated molecular patterns
RIP1	Receptor-interacting protein 1
TABs	TAK1-binding proteins
TAK	TGF-β-activated kinase
TBK1	TRAF family member-associated NF-κB activator binding kinase 1
TGF-β	Transforming growth factor-β
Th2	T-helper 2
TIR	Toll/IL-1 receptor
TIRAP/MAL	TIR-containing adaptor protein/MyD88 adaptor-like
TNF-α	Tumor necrosis factor α
TLRs	Toll-like receptors
TRAF	TNF-associated factor
TRAM/TICAM2	TIR-domain-containing adaptor molecule/TRIF-related adaptor molecule 2
TRIF/TICAM	1 TIR-containing adaptor inducing IFN-β/TIR-domain containing adaptor molecule 1

1 Alzheimer's Disease

Alzheimer's disease (AD) was first described by Dr. Alois Alzheimer in 1907. He described a 51-year-old woman named August D. who exhibited progressive memory loss, cognitive decline, changes in behavior, and loss of language function, which comprise the cardinal symptoms of the disease. AD is the most common cause of dementia in the elderly. Currently, there are 5 million cases in the United States, with over 15 million cases worldwide. The number of AD cases continues to rise as the average life expectancy increases, reaching perhaps 16 million Americans by 2050. Average disease progression occurs over the course of eight years, culminating in death of the patient. It has been argued that if a treatment could delay symptom onset by five years, the number of individuals with AD could ultimately be reduced by nearly 50%. This would relieve both economic and medical burdens, as currently the national total for direct and indirect costs of AD are over $148 billion annually (see http://www.Alz.org).

The AD brain is characterized by extensive extracellular deposits of β-amyloid (Aβ), which condenses to form senile plaques. In addition to amyloid deposits, the AD brain exhibits intracellular neurofibrillary tangles composed of the hyperphosphorylated microtubule-associated protein tau and extensive neuronal atrophy within the hippocampus, and the entorhinal and the temporoparietal cortices. The disease is also characterized by a robust inflammatory response with increased levels of inflammatory cytokines, chemokines, immune cell surface proteins, acute-phase proteins, complement proteins, and oxidative damage within the brain (Akiyama et al. 2000).

Age is the single biggest risk factor for AD; there is a positive correlation between increased age and the incidence of AD (Gorelick 2004). Although most cases are sporadic, a small number of the total cases are inherited and are linked to autosomal dominant mutations in genes whose products participate in amyloid precursor protein (APP) processing and Aβ production (Findeis 2007). People with these mutations have disease onset prior to age 65, whereas individuals affected after the age of 65 have late-onset AD. A number of susceptibility genes have been identified that influence the risk of late-onset AD as well (Findeis 2007). Moreover, vascular, environmental, and social factors in addition to life habits and medication have been shown to be associated with increased or decreased AD risk (Fratiglioni et al. 2007).

The most widely held hypothesis for AD pathogenesis is known as the "amyloid hypothesis." It posits that a gradual and chronic imbalance in the production and clearance of Aβ leads to an increase in its steady-state levels within the brain over the course of decades, thus resulting in the complex molecular and cellular changes within the brain that typify AD (Selkoe 2000). Aβ is produced by the sequential proteolytic processing of APP by either the α- or β-secretase (BACE), and subsequently the γ-secretase. The pathogenic processing is initiated by the cleavage of APP by BACE, releasing APPs-β, leaving the C99 fragment in the membrane. C99 is then cleaved by γ-secretase to produce the Aβ peptide, with the two major forms being 40 and 42 residues in length. In a normal individual, the majority of the Aβ produced is the 40 amino acid species, whereas 5–15% of the total Aβ pool is made up of $A\beta_{42}$ (Younkin 1998). Some Familial APP mutations increase the relative production of the more toxic and more amyloidogenic species, $A\beta_{42}$, which is important in the development of AD (Crouch et al. 2008). In the nonpathogenic APP processing pathway, APP is cleaved by the α-secretase to produce APPs-α and the C83 fragment. C83 is then cleaved by the γ-secretase to produce the nonamyloidogenic p3 fragment.

2 Inflammation and Alzheimer's Disease

Extensive research has demonstrated the involvement of inflammation in AD. There is an extensive literature documenting that the brains and cerebrospinal fluid of AD patients contain a variety of proinflammatory mediators, including complement, cytokines and chemokines, acute-phase proteins, proteases, as well as reactive oxygen and nitrogen species that are responsible for the oxidative damage in the AD brain (reviewed in Akiyama et al. 2000). In addition, a subset of epidemiological studies have shown that chronic use of some nonsteroidal anti-inflammatory drugs (NSAIDs) drastically reduces the risk, onset, severity, and progression of AD (Rogers et al. 1993; Rich et al. 1995; McGeer et al. 1996; Stewart et al. 1997; in't Veld et al. 2000).

The principal immune effector cells of the brain are microglia. A number of hypotheses exist for the origin of microglia; currently, the consensus is that they are derived from mesodermal/mesenchymal sources in the periphery and infiltrate the brain during development (reviewed in Chan et al. 2007). It has recently been suggested that peripherally derived monocytes and macrophages can traffic into and populate the brain, and subsequently perform microglia-related functions (Simard and Rivest 2004). However, this phenomenon has now been shown to be an artifact of the experimental manipulation (reviewed in Ransohoff 2007).

Under normal, nonpathogenic conditions, microglia have a small soma and highly ramified morphology, and exhibit a "resting" morphology, but are far from dormant. Two recent studies have demonstrated the dynamic nature of microglia in immune surveillance of the brain (Davalos et al. 2005; Nimmerjahn et al. 2005). Using the microglial-specific *Cx3cr1* locus to drive enhanced green fluorescent protein (eGFP) expression, two-photon microscopy revealed that microglia are evenly spaced throughout the brain at approximately six cells per cubic millimeter. These cells contain highly mobile processes that continuously extend and retract, resulting in inspection of the entire brain parenchyma every few hours. Neighboring microglial cells alternate in scanning overlapping regions, ensuring thorough examination of the brain parenchyma while avoiding contact. These dynamic and finely tuned surveillance mechanisms enable stationary microglia to scan their environment without disrupting the neuronal network of the brain parenchyma (Hanisch and Kettenmann 2007). Upon recognition of an insult or foreign material, microglia shift their activation state from a surveillance mode to a reactive mode where they act as immune effector cells.

Microglial activation status is determined by their immediate environment (Goerdt and Orfanos 1999; Gordon 2003). The response of microglia, like other tissue macrophages, to cytokines and a proinflammatory environment (Th1 response) results in a classical activation. This activation status is characterized by upregulation of a variety of cell surface receptors, proinflammatory molecules, nitric oxide, and prostaglandins (Mantovani et al. 2002). In contrast, microglia exposed to the anti-inflammatory cytokines interleukin 4 (IL-4), IL-10, IL-13, and transforming growth factor β (TGF-β), reflective of a Th2-type response, become alternatively activated. This alternate activation state demonstrates a greater capacity for phagocytosis and does not produce nitric oxide (Goerdt and Orfanos 1999). A proinflammatory environment reduces the phagocytic capacity of microglia (Koenigsknecht-Talboo and Landreth 2005; Zelcer et al. 2007). In these studies, LPS was used to induce a proinflammatory environment as it stimulates cytokine gene expression via NF-κB. The suppression of phagocytosis by proinflammatory cytokines is reversible by NSAIDs, anti-inflammatory cytokines, and liver X receptor agonists.

Microglia surrounding Aβ plaques show an activated phenotype and extend processes which envelop the Aβ plaque (Bornemann et al. 2001; Bolmont et al. 2008). Compared to those found in the nondemented brain, these microglia express higher levels of a number of cell surface proteins, including the major histocompatibility complex II (MHC II) cell surface glycoproteins (Luber-Narod and Rogers 1988; Rogers et al. 1988; Haga et al. 1989; McGeer et al. 1989; Styren et al. 1990). In addition, a variety of other cell surface proteins are upregulated on activated microglia neighboring Aβ plaques. Microglia surrounding plaques have also been shown to proliferate, contributing to their accumulation at the plaque periphery (Bornemann et al. 2001; Stalder et al. 2001; Bolmont et al. 2008). Some aspects of the microglial inflammatory response represent positive influences with respect to AD pathogenesis, such as phagocytosis, which may participate in the removal of Aβ from the brain. However, decades of direct and bystander damage from a chronic microglia-mediated inflammatory response mechanism likely exacerbates disease pathogenesis. Importantly, the levels of proinflammatory cytokines are dependent on the magnitude of plaque burden in the AD brain (Patel et al. 2005). It has been suggested that the inflammatory response facilitates the production and deposition of Aβ (Akiyama et al. 2000; Patel et al. 2005).

3 Innate Immunity in Alzheimer's Disease

The innate immune system was first described over one hundred years ago by Dr. Elie Metchnikoff. This system mediates the detection of constitutive and conserved products of microbial metabolism and mobilizes a robust immune response resulting in death or disposal of the invading pathogen (reviewed in Medzhitov 2001). Although bacterial products exhibit variation in their chemical structures, they are typically found in the context of a common molecular pattern called pathogen-associated molecular patterns (PAMPs). These PAMPs are recognized by a variety of pattern recognition receptors (PRRs). As PRRs are encoded in the germ line, there is no requirement for previous exposure, so the response is rapid; thus, the innate immune system is the first line of defense in host defense.

Toll-like receptors (TLRs) are a family of PRRs characterized by an extracellular leucine-rich repeat domain and an intracellular Toll/IL-1 receptor (TIR) domain (Kielian 2006). In mammals there are at least ten TLRs; despite a high degree of structural similarity, each receptor has a distinct function in innate immune recognition. Although each TLR has some degree of ligand specificity, this specificity is extended through dimerization of the TLRs. Some TLRs employ additional co-receptors that assist in pathogen recognition, such as CD14 for TLR4 (Kielian 2006).

The binding of ligand to TLRs initiates intracellular signal transduction cascades. The TLR-stimulated signaling cascades act to upregulate the expression of proinflammatory cytokines and chemokines, nitric oxide synthase and other antimicrobial peptides that directly destroy microbial pathogens (reviewed in Han and Ulevitch 2005).

TLRs possess an intracellular Toll/IL-1 receptor (TIR) domain that serves as a scaffold to generate intracellular signaling cascades that induce changes for mechanisms of host defense. Following recognition of ligand, TLRs recruit specific combinations of adaptor proteins that include myeloid differentiation primary response gene 88 (MyD88), TIR-containing adaptor protein/MyD88-adaptor-like (TIRAP/MAL), TIR-containing adaptor inducing interferon-β (IFN-β)/TIR-domain-containing adaptor molecule 1 (TRIF/TICAM1) and TIR-domain-containing adaptor molecule/TRIF-related adaptor molecule 2 (TRAM/TICAM2) (Kawai and Akira 2007). TLR signaling is divided into MyD88-dependent and TRIF-dependent pathways. The MyD88-dependent pathway involves the IL-1 receptor-associated kinase (IRAK) family members IRAK4 and IRAK1, which interact with tumor necrosis factor receptor-associated factor 6 (TRAF6) to promote ubiquitination for recruitment of transforming growth factor β-activated kinase 1 (TAK1) and TAK1-binding proteins (TABs) (Chen et al. 2006a; Adhikari et al. 2007; Kawai and Akira 2007). These molecules then activate pathways involving the IKK complex and the mitogen-activated protein kinase (MAPK) pathway for NF-κB-dependent proinflammatory gene expression (Wang et al. 2001; Sanjo et al. 2003; Shim et al. 2005). In contrast, TRIF-dependent signaling leads to NF-κB activation through TRAF6, which activates TAK1 in a manner similar to that of the MyD88-dependent pathway via receptor-interacting protein 1 (RIP1) (Yamamoto et al. 2002; Sato et al. 2003; Jiang et al. 2004; Meylan et al. 2004; Cusson-Hermance et al. 2005). The TRIF-dependent pathway also activates the IKK-related kinases TRAF family member-associated NF-κB activator binding kinase 1 (TBK1) and IKKi through TRAF3 (Pomerantz and Baltimore 1999; Nomura et al. 2000; Kawai and Akira 2007). These kinases phosphorylate the transcription factors interferon regulatory factor 3 (IRF3) and IRF7, activating them for IFN-β induction (Fitzgerald et al. 2003; Sharma et al. 2003).

3.1 Increased TLR Expression

It has been shown that there are changes in neuroinflammation associated with aging (Letiembre et al. 2007). It has been argued that a proinflammatory status originates as an adaptive mechanism of aging that builds up over time, representing a biological background favoring disease susceptibility. The addition of various risk factors, genetic or environmental, results in the development of overt age-related diseases with an inflammatory pathogenesis such as AD (Franceschi et al. 2000). Age-related changes in the regulation of innate immunity may be of clinical significance for the development of mild cognitive deficits or for predisposition to neurodegenerative diseases in the elderly (Letiembre et al. 2007).

An unexplained feature of normal aging is an increase in innate immune receptor expression in the brains of aging mice (Letiembre et al. 2007). Importantly, there is also increased expression of these receptors in brains of AD patients (Liu et al. 2005; Letiembre et al. 2009; Walter et al. 2007) and in animal models of the disease (Fassbender et al. 2004; Letiembre et al. 2009; Walter et al. 2007). High levels of CD14 were found in the microglia of the cortex and hippocampus of the APP23 mouse model of AD (Fassbender et al. 2004). AD patients showed increased CD14 expression in parenchymal microglia of the frontal and occipital cortex, hippocampus, and around the senile plaques. Some perivascular cells of the brain exhibited CD14 immunoreactivity as well (Liu et al. 2005). The AD mouse model TgCRND8 was reported to have higher TLR4 mRNA expression as compared to age-matched controls, and brains of AD patients exhibited pronounced TLR4 expression by immunofluorescence that was associated with Aβ plaque deposition in the entorhinal cortex (Walter et al. 2007). Infusion of $A\beta_{42}$ into the hippocampus induced TLR2 expression (Richard et al. 2008). A screen of innate immune receptors in the TgCRND8 animal also revealed an upregulation of TLR2 and TLR7 expression (Letiembre et al. 2009). CD14 expression was localized around the Aβ plaques present in the cortex. However, TLR2 exhibited a more complex pattern of expression; it was absent from the cortex but was found surrounding Aβ plaques in the amygdala, where CD14 was largely undetectable. In human AD brains, both TLR2 and CD14-positive microglia were found associated with Aβ plaques, and a greater number of CD14-positive cells are found surrounding diffuse plaques compared to dense-core plaques (Letiembre et al. 2009). These studies suggest that increased expression of CD14 and TLRs is associated with increased inflammation associated with aging, and when combined with various AD risk factors, this inflammatory environment results in exacerbation of disease progression through an inflammatory response.

Interestingly, a recent study of a TLR4 polymorphism that exhibits a blunted TLR signaling response is associated with a 2.7-fold reduction in risk of late-onset AD (Minoretti et al. 2006). An adenine to guanine substitution in TLR4 causes the replacement of an aspartic acid residue by a glycine at amino acid 299. This affects the structure of the extracellular domain of TLR4 and results in attenuated efficacy of LPS signaling and a reduced capacity to elicit inflammation (Arbour et al. 2000). Work by Minoretti et al. demonstrated that the frequency of the 299Gly allele was significantly higher in the nondemented age-matched controls than in the AD cases, suggesting that TLR signaling may play a role in AD pathogenesis (Minoretti et al. 2006). In another study, subjects having a specific combination of polymorphisms in CD14 and LXRβ had a sixfold reduction in the risk of developing AD (Rodriguez-Rodriguez et al. 2008). The polymorphism in the promoter region of CD14 decreases CD14 expression on circulating monocytes and in plasma. This coupled with a polymorphism in LXRβ intron 5 is thought to lower the inflammatory response in the brain, which in turn decreases the risk of AD (Rodriguez-Rodriguez et al. 2008).

3.2 TLRs in Microglial Activation

Only recently have TLRs been implicated in microglial activation in AD, and how microglial TLRs function in the inflammatory response in AD is now under active investigation. TLR ligands were typically thought to be "exogenous," in that they were of microbial origin. However, recent evidence suggests the presence of "endogenous" TLR ligands. Interestingly, the majority of endogenous TLR ligands interact with TLR4 or TLR2 and 4, and not with other members of the TLR family (reviewed in Tsan and Gao 2007). The diversity of TLR ligands has been explained by a hydrophobicity model (Seong and Matzinger 2004), where hydrophobic domains in any molecule can become molecular patterns recognized by TLRs. Because Aβ forms hydrophobic aggregates in AD plaques, it was investigated whether CD14 could mediate Aβ-induced neuroinflammation (Fassbender et al. 2004; Liu et al. 2005). Using a variety of techniques, CD14 was shown to bind $A\beta_{42}$. The interaction between CD14 and fibrillar $A\beta_{42}$ was 20 times greater than that between CD14 and nonfibrillar $A\beta_{42}$, indicating the importance of the fibrillar structure of $A\beta_{42}$ in binding CD14. Although the affinity of the $A\beta_{42}$–CD14 interaction is approximately 50 times lower than that of LPS and CD14, it is thought that because AD brains contain high concentrations of $A\beta_{42}$ over a period of decades, it is likely that even at this submaximal affinity, this interaction is sufficient to maintain chronic neuroinflammation (Fassbender et al. 2004). Fassbender and colleagues concluded that CD14 along with "accessory receptors" may bind highly hydrophobic Aβ aggregates in a manner similar to that of exogenous PAMPs in a "structural mimicry" mechanism (Fassbender et al. 2004). Inhibition of CD14 function, either through the use of a function blocking antibody or the use of CD14-deficient microglia, demonstrated that CD14 was required for Aβ/IFNγ-induced release of nitrite, IL-6, and TNF-α in both murine microglia and in human peripheral blood monocytes. This response was associated with NFκB nuclear translocation and activation (Fassbender et al. 2004).

These studies led to the question of whether TLRs participated in Aβ-induced microglial activation. Fibrillar Aβ engages TLRs to stimulate "host defense" mechanisms, resulting in proinflammatory activation of microglia. Inhibition of TLR4 in human monocytes or murine microglia, through the use of function blocking antibodies, resulted in reduced nitrite, TNF-α, and IL-6 production following aggregated, fibrillar $A\beta_{42}$ exposure (Walter et al. 2007; Udan et al. 2008). Nonfibrillar forms of $A\beta_{42}$ were unable to stimulate IL-8 secretion in HEK293 cells expressing CD14, TLR4, and MD2, again suggesting that the aggregated conformation of Aβ is a structural prerequisite for cellular activation by innate immune receptors (Walter et al. 2007). Microglia without functioning TLR2 were unable to increase expression of TNF-α, IL-1β, IL-6, the integrin markers CD11b, CD11c, and CD68, in addition to inducible NO synthase following fibrillar $A\beta_{42}$ (Jana et al. 2008; Udan et al. 2008). TLR2 expression in HEK293 cells resulted in Aβ-induced IL-8 production, and was enhanced after CD14 co-expression (Walter et al. 2007). Moreover, animal models of AD have shown a requirement for TLRs in microglial activation and inflammatory gene expression. A mouse model of AD lacking functional TLR4 had increased levels of the microglial marker CD11b and the reactive astrocyte marker GFAP (Jin et al. 2008). In addition, AD mice deficient in TLR2 were shown to have increased TGF-β mRNA compared to their TLR2+/+ littermates (Richard et al. 2008).

A microglial receptor complex comprised of the $\alpha_6\beta_1$ integrin, the integrin-associated protein CD47, and CD36 has been described for fibrillar Aβ (Bamberger et al. 2003). This receptor complex is required for Aβ-induced cytokine production, reactive oxygen species production and a phagocytic response (Bamberger et al. 2003; Koenigsknecht and Landreth 2004). Microglia lacking CD14, TLR4, or TLR2 are unable to produce reactive oxygen species or a phagocytic response following stimulation with fibrillar Aβ (Reed-Geaghan and Landreth 2007). Wilkinson et al. demonstrated that following engagement of this receptor complex by fibrillar Aβ, Lyn and Syk kinases associate with Vav, a Rac guanine nucleotide exchange factor, phosphorylating and thereby activating it (Wilkinson et al. 2006). Vav activation permitted Rac participation in NADPH oxidase complex formation for reactive oxygen species production and for participation in actin cytoskeletal rearrangement for phagocytosis. Microglia deficient for CD14, TLR4, or TLR2 were unable to stimulate signaling through this Src–Vav–Rac cascade (Reed-Geaghan and Landreth 2007). The MAPK p38 becomes activated following fibrillar Aβ exposure (McDonald et al. 1998), and has been shown to function in TLR-dependent endosome maturation (Blander and Medzhitov 2004) as well as in phosphorylation of the NADPH oxidase subunit $p47^{phox}$ for reactive oxygen species production (El Benna et al. 1996). Following treatment with fibrillar Aβ, microglia from CD14, TLR4, or TLR2-null animals were unable to activate p38 (Reed-Geaghan and Landreth 2007). These studies demonstrate the involvement of CD14 and the TLRs in microglial recognition of fibrillar Aβ and subsequent signal transduction cascades for microglial activation.

Persistent activation of microglia results in their migration to the activating stimulus. Human G protein-coupled formyl peptide receptor-like 1 (FPRL1) and its mouse homolog murine formyl peptide receptor 2 (mFPR2) mediate the chemotactic activity and uptake of $A\beta_{42}$ by monocytic cells, and thus may be involved in recruiting microglia to amyloid plaques (Tiffany et al. 2001; Yazawa et al. 2001; Cui et al. 2002; Ying et al. 2004; Iribarren et al. 2005b). The TLR ligands LPS, CpG, and PGN upregulate mFPR2 mRNA in both a microglial cell line and in primary murine microglia, accompanied by an increased chemotactic response to Aβ and increased endocytosis of Aβ (Iribarren et al. 2005a; Chen et al. 2006b). Activation of the p38 and ERK1/2 MAP kinases as well as activation of NF-κB by TLR ligands is required for the induction of mFPR2 mRNA and the chemotactic response (Iribarren et al. 2005a; Chen et al. 2006b). Here the activation of TLR4 may promote a microglial response in AD where the mFPR2 agonist Aβ is elevated. Thus, TLRs on the surface of microglia act as sensors for proinflammatory signals and orchestrate the host response in the brain (Iribarren et al. 2005a).

While all previous studies point to TLR ligands enhancing the activation of microglia by Aβ, a recent study by Lotz et al. suggests that not all TLR agonists enhance the stimulatory effect of Aβ on innate immunity (Lotz et al. 2005). In these studies, simultaneous treatment of primary mouse microglia or murine peritoneal macrophages with LPS or the TLR2 ligand Pam3Cys together with $A\beta_{40}$ had an additive effect on nitric oxide and TNF-α release compared to LPS or Pam3Cys alone.

However, simultaneous treatment of microglia or macrophages with the TLR9 ligand single-stranded unmethylated CpG-DNA and Aβ_{40}, while having little effect at low doses, significantly decreased NO release at higher concentrations. Interestingly, low doses of CpG with Aβ_{40} reduced TNF-α release. Treatment of microglia with Aβ_{40} resulted in microglial activation characterized by cell rounding, loss of ramification, and formation of cytoplasmic vacuoles. Aβ was detected both at the surface and within the cell, but did not co-localize with CpG. The authors propose that the distinct responses of microglia to individual TLR agonists is the result of receptor localization and the difference in signaling cascades (Lotz et al. 2005).

3.3 TLRs in the Death of Neurons

Neuronal loss is a prominent element of AD pathology. Combs et al. demonstrated that neuronal apoptosis was induced by the addition of conditioned media from Aβ-stimulated monocytes (Combs et al. 2001). TLRs have been shown to have a role in neuronal apoptosis as well. Hippocampal neurons exposed to conditioned media from microglia treated with aggregated Aβ_{42} resulted in neuronal death. However, media from CD14- or TLR4-knockout microglia treated with Aβ_{42} were unable to kill neurons (Fassbender et al. 2004; Walter et al. 2007). These data suggest that CD14 and TLR4 function in the production of neurotoxic molecules.

Another mechanism whereby TLRs function in neuronal death is through the recognition of Aβ-damaged neurons. Bate et al. have shown that microglia kill Aβ_{42}-damaged neurons through a contact-dependent mechanism (Bate et al. 2004). The addition of microglia to neurons treated with sublethal doses of Aβ_{42} resulted in the death of neurons in a manner inversely related to both the concentration of Aβ and the number of microglial cells added (Bate et al. 2006). Involvement of CD14 in this response was demonstrated by the attenuated killing of Aβ-treated neurons following pretreatment of microglia with an antibody to CD14, as well as the inability of microglia from CD14–/– mice to kill Aβ-treated neurons (Bate et al. 2004). Moreover, neurons treated with Aβ_{42} possess higher levels of CD14-reactive molecules on their cell surface, as they bound higher amounts of a CD14-IgG chimera than did untreated neurons (Bate et al. 2006). Significantly, the addition of this CD14-IgG chimera to Aβ-treated neurons prevented microglial killing of neurons without altering the direct effects of Aβ (Bate et al. 2004). The authors hypothesized that neurons that have survived low (i.e., nontoxic) concentrations of Aβ undergo phenotypic changes which lead to them being recognized and killed by microglial cells via CD14. This model may mimic some of the earliest neuronal changes, such as loss of synapses and axon terminal degeneration, observed during disease progression (Bate et al. 2006).

3.4 TLRs and the Clearance of Aβ from the Brain

The imbalance between Aβ production and clearance influences AD pathogenesis. In vitro, treatment of microglia with the TLR4 ligand LPS or the TLR9 ligand CpG oligodeoxynucleotide stimulates uptake of Aβ (Tahara et al. 2006). Fiala et al. examined the difference in Aβ uptake and TLR expression (Fiala et al. 2007) and found that macrophages from control subjects usually showed extremely efficient phagocytosis of Aβ and rapid intracellular transport of Aβ. In contrast, macrophages from AD patients minimally ingested Aβ and did not transport Aβ into endosomes and lysosomes. Control macrophages stimulated with Aβ increased expression of TLRs, but AD macrophages downregulated their TLR ratios. The authors hypothesized that the lower expression levels of TLRs on AD macrophages may be indicative of more global innate immune defects beyond Aβ phagocytosis, where the innate and adaptive immune systems are in various states of dissonance (Fiala et al. 2007).

Because CD14 has been shown to be required for the uptake of a variety of pathogens, the involvement of CD14 in Aβ internalization has been studied (Liu et al. 2005). Confocal analysis of wild-type microglia showed CD14 and fibrillar $A\beta_{42}$ co-localized at the cell surface, and after 30 min, this complex was internalized and co-localized with the lysosomal marker LAMP2. CD14–/– microglia phagocytosed less fibrillar $A\beta_{42}$ than did their wild-type counterparts, though this was not due to general impairment of the phagocytic machinery. These data clearly demonstrate the requirement for CD14 in microglial phagocytosis of Aβ.

The importance of TLR4 and TLR2 in Aβ uptake has also been assessed in animal models of AD. The APPswe/PSEN1dE9 mouse model of Alzheimer's disease (Jankowsky et al. 2004) has as part of its background the C3H/HeJ strain. These animals contain a co-dominant destructive point mutation in the TLR4 gene, resulting in the failure of TLR4 activation by LPS (Poltorak et al. 1998). Tahara et al. took advantage of this intrinsic mutation to evaluate the role of TLR4 in amyloidogenesis in vivo (Tahara et al. 2006). As compared to their wild-type littermates, APPswe/PSENdE9 mice with inactive TLR4 showed increased cortical and hippocampal Aβ load without changing steady-state APP or presenilin 1 levels. These data strongly argue that the change in Aβ load was due to a change in microglia-mediated Aβ clearance that is reliant upon TLR4 function. APPswe/PSENdE9 mice have also been mated to TLR2–/– mice, and the resulting transgenic TLR2–/– mice showed delayed Aβ deposition through six months of age but had comparable deposition by nine months of age compared to their TLR2+/+ littermates (Richard et al. 2008). The different results for Aβ deposition between TLR4- and TLR2-deficient animals could be explained by the age of the animals assessed. Jin et al. used mice that were aged at least 14 months, while Richard et al. examined their animals no later than nine months of age. The APPswe/PSENdE9 mouse has decreased expression of various Aβ-binding receptors and degrading enzymes at eight months of age, in addition to increased proinflammatory gene expression (Hickman et al. 2008). It is hypothesized that, while microglia clear Aβ in the early stages of AD, as the disease progresses, genes involved in Aβ clearance are downregulated due to increased proinflammatory gene expression, contributing to Aβ accumulation (Hickman et al. 2008). Thus, CD14, TLR4, and TLR2 function in the recognition and binding of Aβ for its internalization and clearance from the brain parenchyma.

4 Conclusions

The AD brain contains many potential inflammatory stimuli, including Aβ and damaged neurons. Due to the presence of these stimuli, there is a discrete, localized inflammatory reaction. Over the course of decades, this chronic inflammation produces direct and indirect damage within the brain. Amyloid deposition begins 10–20 years before the appearance of clinical dementia, allowing inflammatory processes to proceed for years unchecked. A small number of studies prospectively investigated the relationship between systemic markers of inflammation and the risk of cognitive decline (reviewed in Dziedzic 2006). Individuals with high levels of the inflammatory markers C-reactive protein, IL-6, and α1-antichymotrypsin were associated with increased dementia. Work has shown that, to a degree, the microglial inflammatory response has beneficial effects. Microglia migrate to plaques for phagocytosis and degradation of Aβ. They upregulate a variety of factors that aid in the clearance of Aβ, such as complement, integrins, and Fc receptors (reviewed in Blasko and Grubeck-Loebenstein 2003). However, while microglial activation is suppressed by electrically active neurons (Neumann and Wekerle 1998), in AD brain, neurons continuously exposed to Aβ lose the ability to suppress immune responses, so that a chronic inflammatory response exists, thereby contributing to disease pathology.

Work discussed above illustrates the role of CD14 and TLRs in this process. CD14 has the ability to recognize and bind Aβ aggregates, and along with TLRs mediate microglial activation and clearance of Aβ from the brain (Fassbender et al. 2004; Liu et al. 2005; Tahara et al. 2006; Walter et al. 2007; Udan et al. 2008). In addition to CD14 and TLRs functioning in microglial activation, CD14 functions in the recognition and clearance of Aβ-damaged neurons (Bate et al. 2004, 2006). Thus, the innate immune system is truly a double-edged sword. It has been proposed that at low Aβ concentrations corresponding to those observed in the brain of early/middle-stage AD, CD14 and TLRs may activate microglia promoting the phagocytic clearance of Aβ, whereas at higher Aβ concentrations corresponding to those at late-stage AD, microglial activation through CD14 and the TLRs results in the production of neurotoxins as well, thereby damaging surrounding neurons (Liu et al. 2005; Walter et al. 2007) and killing these damaged neurons. The recognition of the involvement of TLRs and their co-receptors in AD pathogenesis suggests that they may be an appropriate target for therapeutic intervention within the disease progression.

Acknowledgments This work was supported by the NIA (AG16047), the Blanchette Hooker Rockefeller Foundation, and the American Health Assistance Foundation. Erin Reed-Geaghan is supported by a predoctoral Ruth L. Kirschstein National Research Service Award (F31NS057867) from the NINDS.

References

Adhikari A, Xu M, Chen Z (2007) Ubiquitin-mediated activation of TAK1 and IKK. Oncogene 26:3214–3226

Akiyama H, Barger S, Barnum S, Bradt B, Bauer J, Cole G, Cooper N, Eikelenboom P, Emmerling M, Fiebich B, Finch C, Frautschy S, Griffin W, Hampel H, Hull M, Landreth G, Lue L, Mrak R, Mackenzie I, McGeer P, O'Banion K, Pachter J, Pasinetti G, Plata-Salaman C, Rogers J, Rydel R, Shen Y, Streit W, Strohmeyer R, Tooyoma I, Van Muiswinkel F, Veerhuis R, Walker D, Webster S, Wegrzyniak B, Wenk G, Wyss-Coray T (2000) Inflammation and Alzheimer's disease. Neurobiol Aging 21:383–421

Arbour N, Lorenz E, Schutte B, Zabner J, Kline J, Jones M, Frees K, Watt J, Schwartz D (2000) TLR4 mutations are associated with endotoxin hyporesponsiveness in humans. Nat Genet 25:187–191

Bamberger ME, Harris ME, McDonald DR, Husemann J, Landreth GE (2003) A cell surface receptor complex for fibrillar beta-amyloid mediates microglial activation. J Neurosci 23:2665–2674

Bate C, Kempster S, Williams A (2006) Prostaglandin D2 mediates neuronal damage by amyloid-beta or prions which activate microglial cells. Neuropharmacology 50:229–237

Bate C, Veerhuis R, Eikelenboom P, Williams A (2004) Microglia kill amyloid-β1–42 damaged neurons by a CD14-dependent process. NeuroReport 15:1427–1430

Blander J, Medzhitov R (2004) Regulation of phagosome maturation by signals from toll-like receptors. Science 304:1014–1018

Blasko I, Grubeck-Loebenstein B (2003) Role of the immune system in the pathogenesis, prevention and treatment of Alzheimer's disease. Drugs Aging 20:101–113

Bolmont T, Haiss F, Eicke D, Radde R, Mathis C, Klunk W, Kohsaka S, Jucker M, Calhoun M (2008) Dynamics of the microglial/amyloid interaction indicate a role in plaque maintenance. J Neurosci 28:4283–4292

Bornemann K, Wiederhold K, Pauli C, Ermini F, Stalder M, Schnell L, Sommer B, Jucker M, Staufenbiel M (2001) Abeta-induced inflammatory processes in microglial cells of APP23 transgenic mice. Am J Pathol 158:63–73

Chan W, Kohsaka S, Rezaie P (2007) The origin and cell lineage of microglia: new concepts. Brain Res Rev 53:344–354

Chen F, Bhatia D, Chang Q, Castranova V (2006a) Finding NEMO by K63-linked polyubiquitin chain. Cell Death Differ 13:1835–1838

Chen K, Iribarren P, Hu J, Chen J, Gong W, Cho E, Lockett S, Dunlop N, Wang J (2006b) Activation of Toll-like receptor 2 on microglia promotes cell uptake of Alzheimer disease-associated amyloid beta peptide. J Biol Chem 281:3651–3659

Combs C, Karlo J, Kao S, Landreth G (2001) Beta-amyloid stimulation of microglia and monocytes results in TNFalpha-dependent expression of inducible nitric oxide synthase and neuronal apoptosis. J Neurosci 21:1179–1188

Crouch P, Harding S, White A, Camakaris J, Bush A, Masters C (2008) Mechanisms of Aβ mediated neurodegeneration in Alzheimer's disease. Int J Biochem Cell Biol 40:181–198

Cui Y, Le Y, Yazawa H, Gong W, Wang J (2002) Potential role for the formyl peptide receptor-like 1 (FPRL1) in inflammatory aspects of Alzheimer's disease. J Leukoc Biol 72:628–635

Cusson-Hermance N, Khurana S, Lee T, Fitzgerald K, Kelliher M (2005) Rip1 mediates the Trif-dependent toll-like receptor 3- and 4-induced NF-{kappa}B activation but does not contribute to interferon regulatory factor 3 activation. J Biol Chem 280:36560–36566

Davalos D, Grutzendler J, Yang G, Kim JV, Zuo Y, Jung S, Littman DR, Dustin ML, Gan WB (2005) ATP mediates rapid microglial response to local brain injury in vivo. Nat Neurosci 8:752–758

Dziedzic T (2006) Systemic inflammatory markers and risk of dementia. Am J Alzheimers Dis Other Demen 21:258–262

El Benna J, Han J, Park J, Schmid E, Ulevitch R, Babior B (1996) Activation of p38 in stimulated human neutrophils: phosphorylation of the oxidase component p47phox by p38 and ERK but not by JNK. Arch Biochem Biophys 334:395–400

Fassbender K, Walter S, Kuhl S, Landmann R, Ishii K, Bertsch T, Stalder A, Muehlhauser F, Liu Y, Ulmer A, Rivest S, Lentschat A, Gulbins E, Jucker M, Stafenbiel M, Brechtel K, Walter J, Multhaup G, Penke B, Adachi Y, Hartmann T, Beyreuther K (2004) The LPS receptor (CD14) links innate immunity with Alzheimer's disease. FASEB J 18:203–205

Fiala M, Liu P, Espinosa-Jeffery A, Rosenthal M, Bernard G, Ringman J, Sayre M, Zhang L, Zaghi J, Dejbakhsh S, Chiang B, Hui J, Mahanian M, Baghaee A, Hong P, Cashman J (2007) Innate immunity and transcription of MGAT-III and Toll-like receptors in Alzheimer's disease patients are improved by bisdemethoxycurcumin. Proc Natl Acad Sci USA 104:12849–12854

Findeis M (2007) The role of amyloid β peptide 42 in Alzheimer's disease. Pharmacol Ther 116:266–286

Fitzgerald K, McWhirter S, Faia K, Rowe D, Latz E, Golenbock D, Coyle A, Liao S, Maniatis T (2003) IKKepsilon and TBK1 are essential components of the IRF3 signaling pathway. Nat Immunol 4:491–496

Franceschi C, Bonafe M, Valensin S, Olivieri F, De Luca M, Ottaviani E, De Benedictis G (2000) Inflamm-aging. An evolutionary perspective on immunosenescence. Ann NY Acad Sci 908:244–254

Fratiglioni L, Winblad B, von Strauss E (2007) Prevention of Alzheimer's disease and dementia. Major findings from the Kungsholmen Project. Physiol Behav 92:98–104

Goerdt S, Orfanos CE (1999) Other functions, other genes: alternative activation of antigen-presenting cells. Immunity 10:137–142

Gordon S (2003) Alternative activation of macrophages. Nat Rev Immunol 3:23–35

Gorelick P (2004) Risk factors for vascular dementia and Alzheimer's disease. Stroke 35:2620–2622

Haga S, Akai K, Ishii T (1989) Demonstration of microglial cells in and around senile (neuritic) plaques in the Alzheimer brain. An immunohistochemical study using a novel monoclonal antibody. Acta Neuropathol (Berl) 77:569–575

Han J, Ulevitch R (2005) Limiting inflammatory responses during activation of innate immunity. Nat Immunol 6:1198–1205

Hanisch U, Kettenmann H (2007) Microglia: active sensor and versatile effector cells in the normal and pathologic brain. Nat Neurosci 10:1387–1394

Hickman SE, Allison EK, El Khoury J (2008) Microglial dysfunction and defective beta-amyloid clearance pathways in aging Alzheimer's disease mice. J Neurosci 28:8354–8360

in't Veld B, Ruitenberg A, Launer J (2000) Duration of nonsteroidal anti-inflammatory drug use and risk of Alzheimer's disease. The Rotterdam study. Neurobiol Aging 21:S204

Iribarren P, Chen K, Hu J, Gong W, Cho E, Lockett S, Uranchimeg B, Wang J (2005a) CpG-containing oligodeoxynucleotide promotes microglial cell uptake of amyloid beta 1–42 peptide by up-regulating the expression of the G-protein-coupled receptor mFPR2. FASEB J 19:2032–2034

Iribarren P, Zhou Y, Hu J, Le Y, Wang J (2005b) The role of formyl peptide receptor like 1 (FPRL1/MFPR2) in mononuclear phagocyte responses in Alzheimer's disease. Immunol Res 31:165–176

Jana M, Palencia CA, Pahan K (2008) Fibrillar amyloid-beta peptides activate microglia via TLR2: implications for Alzheimer's disease. J Immunol 181:7254–7262

Jankowsky J, Fadale D, Anderson J, Xu G, Gonzales V, Jenkins N, Copeland N, Lee M, Younkin L, Wagner S, Younkin S, Borchelt D (2004) Mutant presenilins specifically elevate levels of the 42 residue beta-amyloid peptide in vivo: evidence for augmentation of a 42-specific gamma secretase. Human Mol Genet 13:159–170

Jiang Z, Mak T, Sen G, Li X (2004) Toll-like receptor 3-mediated activation of NF-kappaB and IRF3 diverges at Toll-IL-1 receptor domain-containing adapter inducing IFN-beta. Proc Natl Acad Sci USA 101:3533–3538

Jin JJ, Kim HD, Maxwell JA, Li L, Fukuchi K (2008) Toll-like receptor 4-dependent upregulation of cytokines in a transgenic mouse model of Alzheimer's disease. J Neuroinflammation 5:23

Kawai T, Akira S (2007) Signaling to NF-kB by Toll-like receptors. Trends Mol Med 13:460–469

Kielian T (2006) Toll-like receptors in central nervous system glial inflammation and homeostasis. J Neurosci Res 83:711–730

Koenigsknecht J, Landreth G (2004) Microglial phagocytosis of fibrillar beta-amyloid through a beta1 integrin-dependent mechanism. J Neurosci 24:9838–9846

Koenigsknecht-Talboo J, Landreth GE (2005) Microglial phagocytosis induced by fibrillar beta-amyloid and IgGs are differentially regulated by pro-inflammatory cytokines. J Neurosci 25:8240–8249

Letiembre M, Hao W, Liu Y, Walter S, Mihaljevic I, Rivest S, Hartmann T, Fassbender K (2007) Innate immune receptor expression in normal brain aging. Neuroscience 146:248–254

Letiembre M, Liu Y, Walter S, Hao W, Pfander T, Wrede A, Schulz-Schaeffer W, Fassbender K (2009) Screening of innate immune receptors in neurodegenerative diseases: a similar pattern. Neurobiol Aging 30:759–768

Liu Y, Walter S, Stagi M, Cherny D, Letiembre M, Schulz-Schaeffer W, Heine H, Penke B, Neumann H, Fassbender K (2005) LPS receptor (CD14): a receptor for phagocytosis of Alzheimer's amyloid peptide. Brain 128:1778–1789

Lotz M, Ebert S, Esselmann H, Iliev A, Prinz M, Wiazewicz N, Wiltfang J, Gerber J, Nau R (2005) Amyloid beta peptide 1–40 enhances action of Toll like receptor-2 and -4 agonists but antagonizes Toll-like receptor-9-induced inflammation in primary mouse microglial cell cultures. J Neurochem 94:289–298

Luber-Narod J, Rogers J (1988) Immune system associated antigens expressed by cells of the human central nervous system. Neurosci Lett 94:17–22

Mantovani A, Sozzani S, Locati M, Allavena P, Sica A (2002) Macrophage polarization: tumor-associated macrophages as a paradigm for polarized M2 mononuclear phagocytes. Trends Immunol 23:549–555

McDonald DR, Bamberger ME, Combs CK, Landreth GE (1998) β-Amyloid fibrils activate parallel mitogen-activated protein kinase pathways in microglia and THP-1 monocytes. J Neurosci 18:4451–4460

McGeer P, Akiyama H, Itagaki S, McGeer E (1989) Immune system response in Alzheimer's disease. Can J Neurol Sci 16:516–527

McGeer P, Schulzer M, McGeer E (1996) Arthritis and antiinflammatory agents as possible protective factors for Alzheimer's disease: a review of 17 epidemiological studies. Neurology 47:425–432

Medzhitov R (2001) Toll-like receptors and innate immunity. Nat Rev Immunol 1:135–145

Meylan E, Burns K, Hofmann K, Blancheteau V, Martinon F, Kelliher M, Tschopp J (2004) RIP1 is an essential mediator of Toll-like receptor 3-induced NF-kappa B activation. Nat Immunol 5:503–507

Minoretti P, Gazzaruso C, Vito C, Emanuele E, Bianchi M, Coen E, Reino M, Geroldi D (2006) Effect of the functional toll-like receptor 4 Asp299Gly polymorphism on susceptibility of late-onset Alzheimer's disease. Neurosci Lett 391:147–149

Neumann H, Wekerle H (1998) Neuronal control of the immune response in the central nervous system: linking brain immunity to neurodegeneration. J Neuropathol Exp Neurol 57:1–9

Nimmerjahn A, Kirchhoff F, Helmchen F (2005) Resting microglial cells are highly dynamic surveillants of brain parenchyma in vivo. Science 308:1314–1318

Nomura F, Kawai T, Nakanishi K, Akira S (2000) NF-kappaB activation through IKK-i-dependent I-TRAF/TANK phosphorylation. Genes Cells 5:191–202

Patel N, Paris D, Mathura V, Quadros A, Crawford F, Mullan M (2005) Inflammatory cytokine levels correlate with amyloid load in transgenic mouse models of Alzheimer's disease. J Neuroinflammation 2:9

Poltorak A, He X, Smirnova I, Liu M, Van Huffel C, Du X, Birdwell D, Alejos E, Silva M, Galanos C, Freudenberg M, Ricciari-Castagnoli P, Layton B, Beutler B (1998) Defective LPS signaling in C3H/HeJ and C57BL/10ScCr mice: mutations in TLR4 gene. Science 282:2085–2088

Pomerantz J, Baltimore D (1999) NF-kappaB activation by a signaling complex containing TRAF2, TANK and TBK1, a novel IKK-related kinase. EMBO J 18:6694–6704

Ransohoff R (2007) Microgliosis: the questions shape the answers. Nat Neurosci 10:1507–1509

Reed-Geaghan E, Landreth G (2007) CD14 and its associated toll like receptors mediate microglial activation by fAβ. In: Society for Neuroscience (ed) 2007 Neuroscience Meeting Planner, vol 688. Society for Neuroscience, San Diego

Rich JB, Rasmusson DX, Folstein MF, Carson KA, Kawas C, Brandt J (1995) Nonsteroidal anti-inflammatory drugs in Alzheimer's disease. Neurology 45:51–55

Richard KL, Filali M, Prefontaine P, Rivest S (2008) Toll-like receptor 2 acts as a natural innate immune receptor to clear amyloid beta 1–42 and delay the cognitive decline in a mouse model of Alzheimer's disease. J Neurosci 28:5784–5793

Rodriguez-Rodriguez E, Sanchez-Juan P, Mateo I, Infante J, Sanchez-Quintana C, Garcia-Gorostiaga I, Berciano J, Combarros O (2008) Interaction between CD14 and LXRbeta genes modulates Alzheimer's disease risk. J Neurol Sci 264:97–99

Rogers J, Kirby LC, Hempelman SR, Berry DL, McGeer PL, Kaszniak AW, Zalinski J, Cofield M, Mansukhani L, Willson P, et al. (1993) Clinical trial of indomethacin in Alzheimer's disease. Neurology 43:1609–1611

Rogers J, Luber-Narod J, Styren S, Civin W (1988) Expression of immune system-associated antigens by cells of the human central nervous system: relationship to the pathology of Alzheimer's disease. Neurobiol Aging 9:339–349

Sanjo H, Takeda K, Tsujimura T, Ninomiya-Tsuji J, Matsumoto K, Akira S (2003) TAB2 is essential for prevention of apoptosis in fetal liver but not for interleukin-1 signaling. Mol Cell Biol 23:1231–1238

Sato S, Sugiyama M, Yamamoto M, Watanabe Y, Kawai T, Takeda K, Akira S (2003) Toll/IL-1 receptor domain-containing adaptor inducing IFN-beta (TRIF) associates with TNF receptor-associated factor 6 and TANK-binding kinase 1, and activates two distinct transcription factors, NF-kappa B and IFN-regulatory factor-3, in the Toll-like receptor signaling. J Immunol 171:4303–4310

Selkoe D (2000) Toward a comprehensive theory for Alzheimer's disease. Hypothesis: Alzheimer's disease is caused by the cerebral accumulation and cytotoxicity of amyloid beta-protein. Ann NY Acad Sci 924:17–25

Seong S, Matzinger P (2004) Hydrophobicity: an ancient damage-associated molecular pattern that initiates innate immune responses. Nat Rev 4:469–478

Sharma S, tenOever B, Grandvaux N, Zhou G, Lin R, Hiscott J (2003) Triggering the interferon antiviral response through an IKK-related pathway. Science 300:1148–1151

Shim J, Xiao C, Paschal A, Bailey S, Rao P, Hayden M, Lee K, Bussey C, Steckel M, Tanaka N, Yamada G, Akira S, Matsumoto K, Ghosh S (2005) TAK1, but not TAB1 or TAB2, plays an essential role in multiple signaling pathways in vivo. Genes Dev 19:2668–2681

Simard A, Rivest S (2004) Bone marrow stem cells have the ability to populate the entire central nervous system into fully differentiated parenchymal microglia. FASEB J 18:998–1000

Stalder M, Deller T, Staufenbiel M, Jucker M (2001) 3D-reconstruction of microglia and amyloid in APP23 transgenic mice: no evidence of intracellular amyloid. Neurobiol Aging 22:427–434

Stewart W, Kawas C, Corrada M, Metter E (1997) Rish of Alzheimer's disease and duration of NSAID use. Neurology 4:626–632

Styren S, Civin W, Rogers J (1990) Molecular, cellular, and pathologic characterization of HLA-DR immunoreactivity in normal elderly and Alzheimer's disease brain. Exp Neurol 110:93–104

Tahara K, Kim H, Jin J, Maxwell J, Li L, Fukuchi K (2006) Role of toll-like receptor signalling in Abeta uptake and clearance. Brain 129:3006–3019

Tiffany H, Lavigne M, Cui Y, Wang J, Leto T, Gao J, Murphy M (2001) Amyloid-beta induces chemotaxis and oxidant stress by acting at formylpeptide receptor 2, a G protein-coupled receptor expressed in phagocytes and brain. J Biol Chem 276:645–652

Tsan M, Gao B (2007) Pathogen-associated molecular pattern contamination as putative endogenous ligands of Toll-like receptors. J Endotoxin Res 13:6–14

Udan M, Ajit D, Crouse N, Nichols M (2008) Toll-like receptors 2 and 4 mediate Aβ(1–42) activation of the innate immune response in a human monocytic cell line. J Neurochem 104:524–533

Walter S, Letiembre M, Liu Y, Heine H, Penke B, Hao W, Bode B, Manietta N, Walter J, Schulz-Schaeffer W, Fassbender K (2007) Role of the Toll-like receptor 4 in neuro-inflammation in Alzheimer's disease. Cell Physiol Biochem 20:947–956

Wang C, Deng L, Hong M, Akkaraju G, Inoue J, Chen Z (2001) TAK1 is a ubiquitin-dependent kinase of MKK and IKK. Nature 412:346–351

Wilkinson B, Koenigsknecht-Talboo J, Grommes C, Lee CY, Landreth G (2006) Fibrillar beta-amyloid-stimulated intracellular signaling cascades require Vav for induction of respiratory burst and phagocytosis in monocytes and microglia. J Biol Chem 281:20842–20850

Yamamoto M, Sato S, Mori K, Hoshino K, Takeuchi O, Takeda K, Akira S (2002) A novel Toll/IL-1 receptor domain-containing adapter that preferentially activates the IFN-beta promoter in the Toll-like receptor signaling. J Immunol 169:6668–6672

Yazawa H, Yu Z, Takeda X, Le Y, Gong W, Ferrans V, Oppenheim J, Li C, Wang J (2001) Beta amyloid peptide (Abeta42) is internalized via the G-protein-coupled receptor FPRL1 and forms fibrillar aggregates in macrophages. FASEB J 15:2454–2462

Ying G, Iribarren P, Zhou Y, Gong W, Zhang N, Yu Z, Le Y, Cui Y, Wang J (2004) Humanin, a newly identified neuroprotective factor, uses the G protein coupled formylpeptide receptor-like-1 as a functional receptor. J Immunol 172:7078–7085

Younkin S (1998) The role of A beta 42 in Alzheimer's disease. J Physiol Paris 92:289–292

Zelcer N, Khanlou N, Clare R, Jiang Q, Reed-Geaghan E, Landreth G, Vinters H, Tontonoz P (2007) Attenuation of neuroinflammation and Alzheimer's disease pathology by liver x receptors. Proc Natl Acad Sci USA 104:10601–10606

Toll-Like Receptors in Multiple Sclerosis

Michael K. Racke and Paul D. Drew

Contents

Abstract Multiple sclerosis (MS) is a chronic disease of the central nervous system (CNS) characterized by inflammation, demyelination, and axonal pathology. The exact causes of MS are unknown, but environmental factors including pathogens are believed to contribute to the development of disease. Toll-like receptors (TLRs) are a family of receptors important in pathogen recognition and host defense. TLRs are expressed by a variety of peripheral immune cells as well as resident cells of the CNS. Studies indicate that TLRs play a significant role in modulating MS, as well as experimental autoimmune encephalomyelitis (EAE), an animal model of MS. This review will discuss the current understanding of the role of TLRs in modulating EAE and MS.

M.K. Racke
Department of Neurology and Department of Molecular Virology, Immunology & Medical Genetics, The Ohio State University Medical Center, 1654 Upham Drive, 445 Means Hall, Columbus, OH, 43210, USA

P.D. Drew (✉)
Department of Neurobiology and Developmental Sciences, University of Arkansas for Medical Sciences, Little Rock, AR, 72205, USA
e-mail: DrewPaulD@uams.edu

T. Kielian (ed.), *Toll-like Receptors: Roles in Infection and Neuropathology*,
Current Topics in Microbiology and Immunology 336,
DOI 10.1007/978-3-642-00549-7_9, © Springer-Verlag Berlin Heidelberg 2009

Abbreviations

APC	Antigen-presenting cell
CFA	Complete Freund's adjuvant
CNS	Central nervous system
EAE	Experimental autoimmune encephalomyelitis
I-κB	Inhibitor κB
IFN	Interferon
IL	Interleukin
IRAK	Interleukin-1 receptor-associated kinase
MS	Multiple sclerosis
MyD88	Myeloid differentiation primary response protein 88
NF-κB	Nuclear factor κB
PAMP	Pathogen-associated molecular patterns
PGN	Peptidoglycan
PRR	Pattern recognition receptor
Th	T helper
TLR	Toll-like receptor
TRAM	TRIF-related adaptor molecule
TRIF	TIR-domain-containing adaptor inducing interferon β

1 Multiple Sclerosis and Experimental Autoimmune Encephalomyelitis

Multiple sclerosis (MS) is a chronic disease of the central nervous system (CNS) characterized by demyelination and progressive axonal degeneration. MS is the second leading cause of neurologic disability in young adults following trauma (Anderson et al. 1992). Most patients suffer from the physical, psychological, and financial effects of MS for most of their adult life. In recent years, a variety of therapeutic agents have been developed which slow disease progression (Johnson et al. 1995; Jacobs et al. 1996; Millefiorini et al. 1997). However, there remains no cure for the disease, and more effective therapies are desperately needed. The cause of MS remains unknown; however, an autoimmune process is hypothesized to be involved in disease pathogenesis (Arnason 1983; Martin et al. 1992; McFarland and Martin 2007). Epidemiologic reports and studies examining the disease in identical twins also suggest that both environment and genetics influence the onset and pathogenesis of MS (Sadovnick and Ebers 1993). MS is believed to be principally mediated by $CD4^+$ T cells that are reactive against myelin antigens (Frohman et al. 2006). These cells are activated in the periphery and express adhesion molecules which facilitate interactions with ligands present on vascular endothelial cells, resulting in extravasation across the blood-brain barrier (Compston 2004). Once in the CNS, these myelin-reactive $CD4^+$ T cells contribute to the demyelination and progressive axonal pathology characteristic of MS (Frohman et al. 2005). In addition to $CD4^+$ T cells, $CD8^+$ T cells and B cells are believed to contribute to the pathogenesis associated with MS (Johnson et al. 2007; Nikbin et al. 2007; Hauser et al. 2008). In addition to lymphocytes that control adaptive immune responses, dendritic cells and tissue macrophages that regulate innate immune responses also play a role in controlling MS disease pathogenesis. These cells express pattern recognition receptors (PRRs) including Toll-like receptors (TLRs) that recognize pathogen-associated molecular patterns (PAMPs) present on the surface of pathogens. Following ligand binding to TLRs, innate immune cells produce proinflammatory cytokines and can serve as antigen-presenting cells (APCs) to prime naïve T cells to recognize antigens (Takeda and Akira 2005; Hacker et al. 2006). Thus, TLRs play an important role in linking the innate to the adaptive immune response (Bell et al. 2005). Finally, glia including astrocytes and microglia play an important role in protecting the CNS against pathogenic insults. However, when chronically activated, these glia may contribute to the pathogenesis of MS, and this may occur in part through PAMP binding to TLRs present on these cells, which can contribute to the reactivation of myelin-specific autoreactive T cells in the CNS (Sanders and De Keyser 2007; Nair et al. 2008). Interestingly, a variety of resident cells of the CNS express TLRs (Bsibsi et al. 2002). Depending on the TLR evaluated, TLR expression on CNS cells has been demonstrated to contribute to oligodendrocyte and neuron cell death (Lehnardt et al. 2002, 2003) or alternatively to be neuroprotective (Bsibsi et al. 2006). It is also interesting that TLRs are expressed on B cells and T cells and that TLR signaling can directly alter adaptive immune responses (Kabelitz 2007; Lampropoulou et al. 2008).

Several animal models have been used to study MS. In some of these models, the disease is induced by viruses, such as Theiler's virus or Borna disease virus (Miller and Karpus 1994). However, the most common model of MS is termed experimental autoimmune encephalomyelitis (EAE). Active EAE is elicited by immunization with a variety of myelin peptides into organisms including rodents and monkeys. EAE can also be elicited following adoptive transfer of myelin-specific T cells into naïve recipients. Of the EAE models, the most commonly studied are those established in the Lewis rat and in several susceptible mouse strains. Murine models of EAE present several advantages over rat models of EAE (Racke 2001). For example, murine EAE results in a relapsing-remitting disease, similar to the early phase of disease for most MS patients, whereas EAE in the Lewis rat is a monophasic illness. In chronic murine EAE, the pathology observed in the white matter shows much more demyelination than the Lewis rat model, again being more reminiscent of the pathology seen in the CNS of patients with MS. In fact, similarities in pathology between mouse EAE and MS have suggested that autoimmunity plays a role in the development of MS. Finally, with the advent of transgenic and homologous recombination technology, it is increasingly clear that many powerful molecular tools are becoming available to study the immune response in pathologic processes such as murine EAE.

2 Role of Pathogens in Multiple Sclerosis and Experimental Autoimmune Encephalomyelitis

The cause of MS is unknown. However, clinical evidence suggests an autoimmune response directed against myelin, possibly stimulated by an infectious agent (Martin et al. 1992; Noseworthy 1999). Epidemiologic studies also suggest that MS is triggered by environmental factors, and an infectious agent is the most likely culprit (Kurtzke 1993). Infections can exacerbate the course of MS, but many different viruses and bacteria have been evaluated (Sibley et al. 1985; Compston et al. 1986; Gay et al. 1986; Panitch 1994; Rapp et al. 1995; Sriram et al. 1998; Horwitz and Sarvetnick 1999; Lenz et al. 2001; Du et al. 2002) and no single environmental agent has been definitively linked to MS development. Furthermore, the mechanisms by which pathogens precipitate MS have not been elucidated. One possibility is that antigens from infectious pathogens may activate autoreactive T cells, causing their expansion and leading to clinical disease (Oldstone 1987; Hausmann and Wucherpfennig 1997). However, molecular mimicry—as this process is termed—has found limited experimental support as a cause of autoimmune diseases. Alternatively, MS may be triggered not by a specific pathogen but instead by general infectious processes. This phenomenon, termed bystander activation, is supported by studies indicating that Theiler's virus can activate APCs and consequently activate autoreactive T cells that do not cross-react with Theiler's virus antigens (Miller et al. 1997; Katz-Levy et al. 1999). Bystander activation has been suggested to be initiated by pathogens that trigger immune processes in the CNS, which is likely mediated by TLRs. Tissue damage then unmasks myelin antigens, resulting in epitope spread and the development of autoimmunity (Vanderlugt and Miller 2002; McMahon et al. 2005; Bailey et al. 2007).

A common clinical occurrence in MS patients is the development of changes in neurological function that appear to be due to MS disease activity but are actually the result of a physiological change such as a urinary tract infection, fever, or electrolyte abnormality. These pseudoexacerbations are usually characterized by a recurrence of old symptoms for short durations, a definable metabolic change, and the disappearance of clinical signs once the metabolic change has been corrected (Tauber et al. 2007). For example, a patient with a prior history of leg weakness can become paraplegic by the fever accompanying a urinary tract infection. Dogma suggests that the reason for the pseudoexacerbation involves elevated temperature producing reversible conduction block in the demyelinated axon. It is this conduction block that produces the recurrence of the old symptoms, in this case paraplegia. If truly due to the above described physiology, most pseudoexcerbations should resolve once the infection is resolved and the fever reduced. However, many MS patients that experience a pseudoexacerbation with a urinary tract infection still require steroids to recover from their symptoms and often never recover completely. These studies suggest that bacterial agents such as *Escherichia coli*, which commonly cause urinary tract infections in many MS patients, may stimulate the innate immune response, which subsequently leads to the activation of autoreactive

T cells, thus contributing to the damage that occurs in the CNS during MS exacerbations.

A variety of studies suggest a critical role for pathogens in the development of EAE. Experiments in the 1970s from pioneers in the field of EAE suggested that both mycobacterial components and pertussis toxin were required for disease development following immunization with myelin antigens (Lublin 1982). The McFarlin laboratory demonstrated that EAE could be induced without pertussis toxin using multiple injections of myelin components in complete Freund's adjuvant (CFA) (Brown and McFarlin 1981). More recently, it has become clear that other adjuvants such as CpG DNA, a ligand for TLR9, can also stimulate induction of EAE (Segal et al. 2000; Deng et al. 2003). Interestingly, adjuvants by themselves have limited capacity to induce EAE without concurrent exposure to myelin antigens. An exception occurs with myelin basic protein-specific TCR transgenic mouse, where immunization with CFA alone resulted in the development of EAE (Goverman et al. 1993). In these mice, because of the high frequency of autoreactive T cells, antigen was not required to expand the encephalitogenic T cell pool. Collectively, these studies suggest that pathogens can initiate an encephalitogenic response that contributes to the development of EAE and MS.

3 Toll-Like Receptors: Role in Multiple Sclerosis and Experimental Autoimmune Encephalomyelitis

3.1 TLRs in MS and EAE: An Overview

Innate immunity represents the first line of defense against pathogens and is principally controlled by phagocytic cells and APCs present in the periphery and in the CNS. Innate immunity is capable of responding rapidly to pathogens, but produces no long-term memory response to a given pathogen. Innate immunity functions through germline-encoded receptors termed PRRs, which include TLRs. Mammalian TLRs were identified based on sequence homology to the *Drosophila* Toll protein (Brennan and Anderson 2004), and collectively these TLRs recognize PAMPs common to a variety of bacteria, viruses, and fungi. Engagement of TLRs with PAMPs leads to the pathogen-induced release of proinflammatory molecules by cells of the innate immune system.

Ligand binding to TLRs triggers receptor dimerization. TLRs are capable of interacting with a series of adaptor proteins that mediate different signaling pathways. Myeloid differentiation primary response protein 88 (MyD88) is the most widely utilized TLR adaptor protein and mediates signaling through all TLRs except TLR3 (Medzhitov et al. 1998; Muzio et al. 1998; Takeda and Akira 2005). MyD88 interacts with the threonine-serine kinase interleukin (IL)-1 receptor-associated kinase 4 (IRAK4), which upon activation phosphorylates IRAK1 (Cao et al. 1996a; Yamin and Miller 1997; Li et al. 2002). Subsequently, the IRAKs recruit the ubiquitin ligase tumor necrosis factor receptor-associated factor 6 (TRAF-6), which polyubiquitinates and activates TAK1 kinase (Cao et al. 1996b). TAK1 kinase activates the IKK complex that triggers the proteolytic degradation of inhibitor κB (I-κB), the inhibitor of nuclear factor κB (NF-κB), which unmasks the nuclear localization signal of NF-κB allowing translocation of this transcription complex from the cytoplasm to the nucleus and activation of a wide variety of NF-κB responsive genes, including genes encoding proinflammatory cytokines and co-stimulatory molecules required for activation of the adaptive immune response. In addition to NF-κB activation, MyD88-dependent signaling results in the activation of transcription factors including AP-1 through a MAPK signaling pathway (Takeda and Akira 2005). Studies involving MyD88-deficiency indicate a critical role of this adaptor protein in response to stimulation through numerous TLRs, including TLR2, TLR4, TLR5, TLR7 and TLR9 (Kawai et al. 1999; Hacker et al. 2000; Schnare et al. 2000; Takeuchi et al. 2000; Hayashi et al. 2001; Hemmi et al. 2002; Takeda et al. 2003).

Ligand binding to TLR3 and TLR4 can stimulate MyD88-independent signaling pathways. These pathways result in the activation of the transcription factor IRF3 and late-phase NF-κB and stimulate the production of genes encoding interferon (IFN)-β and co-stimulatory molecules (Kawai et al. 2001). The adaptor molecule TRIF plays a critical role in signaling through TLR3, while the adaptors TRIF and TRAM are required for MyD88-independent signaling through TLR4 (Fitzgerald et al. 2003b; Oshiumi et al. 2003; Yamamoto et al. 2003; Rowe et al. 2006). Interestingly, the noncanonical I-κB kinases IKKε and TBK1 are required for activation of IRF3 and NF-κB in MyD88-independent signaling pathways (Fitzgerald et al. 2003a).

Adaptive immunity is principally mediated by T and B lymphocytes. These cells are capable of recognizing pathogen-associated antigens in a highly specific manner and provide long-term protection to pathogens. Myelin-reactive T cells are believed to modulate the development of MS and EAE. These myelin-reactive T cells activated in the periphery are capable of moving into the CNS, where they destroy myelin-producing oligodendrocytes and also elicit axonal transaction and neurodegeneration. The role of adaptive immunity in MS has been extensively studied. More recently, the role of the innate immune response in MS has begun to be appreciated. Importantly, it is now recognized that the innate immune response, through the production of proinflammatory cytokines and through antigen presentation, plays a critical role in the activation of myelin-specific autoreactive T cells, thus linking the innate and the adaptive immune system in MS pathogenesis.

Myelin-specific autoreactive $CD4^+$ T cells are believed to contribute to pathogenesis in EAE and MS. $CD4^+$ T cells were originally divided into T helper (Th)1 and Th2 subsets, each exhibiting a distinct effector phenotype and cytokine expression profile (Mosmann and Coffman 1989). Th1 cells produce lymphotoxin and IFN-γ but little IL-4, while Th2 cells produce IL-4, IL-5, IL-13, and IL-25. Th1 cells are believed to contribute to the development of EAE (Ando et al. 1989), while Th2 cells are believed to suppress EAE (Kennedy et al. 1992; Khoruts et al. 1995; Rocken et al. 1996). More recently, Th17 cells that express IL-17, IL-21, and IL-22 have been identified as a third subset of CD4$^+$ T cells that are important in modulating EAE (Bettelli et al. 2008). Interestingly, cytokines produced by cells of the innate immune system modulate the differentiation and function of CD4$^+$ T cells. This provides a critical link between TLR signaling (innate immunity) and autoreactive T cells (adaptive immunity) that cooperatively modulate the pathogenesis associated with MS and EAE.

3.2 TLRs in MS and EAE: The Current Understanding

Previous studies suggest that TLR signaling through the MyD88-dependent pathway plays a significant role in regulating the development of EAE. MyD88 signaling results in the activation of transcription factors including NF-κB and AP-1, which activate the expression of a variety of genes encoding proinflammatory cytokines and chemokines, as well as molecules important in antigen presentation. Each of the TLRs evaluated except TLR3 are capable of triggering MyD88-dependent signaling. However, TLR4 is capable of triggering MyD88-dependent as well as MyD88-independent signaling. EAE is commonly induced in susceptible mouse and rat strains by immunization with myelin peptides emulsified in CFA. In addition, pertussis toxin is generally administered to animals at the time of immunization, and often a booster of pertussis toxin is administered two days following immunization (Racke 2001). Importantly, *Mycobacterium tuberculosis* present in CFA is believed to activate a variety of TLRs, including TLR1, TLR2, and TLR4 (Hansen et al. 2006). Mice immunized with myelin peptides in the presence of incomplete Freund's adjuvant (IFA) do not develop EAE, but do develop disease when *M. tuberculosis* is added to the adjuvant (Hansen et al. 2006). In addition, it has been appreciated for some time that the TLR9 agonist CpG ODN emulsified in IFA is capable of inducing EAE following immunization with myelin peptides in rodent strains susceptible to EAE (Segal et al. 2000; Deng et al. 2003; Hansen et al. 2006).

Several studies support a role of TLR2 in modulating EAE. TLR2 serves as a ligand for Gram-positive bacteria including *Staphylococcus aureus* and *Streptococcus pneumoniae*. Molecules present in the cell wall of these organisms, such as peptidoglycan (PGN), serve as PAMPs that are capable of activating TLR2. *S. aureus* PGN added to IFA was demonstrated to stimulate the development of EAE in C57BL/6 mice (Visser et al. 2005). Associated in vitro studies indicated that PGN stimulated the maturation of dendritic cells and antigen uptake by these cells. Furthermore, PGN pulsed dendritic cells were demonstrated to stimulate T cell proliferation and drive T cell differentiation toward a Th1 phenotype believed to facilitate development of EAE. These studies suggest that PGN signaling through TLR2 may stimulate dendritic cell maturation, antigen presentation, and production of effector molecules resulting in Th1 cell differentiation and development of EAE. The Laman laboratory also demonstrated that PGN is observed in association with APCs in the CNS of MS patients as well as nonhuman primates (Visser et al. 2006).

Furthermore, PGN-laden APCs were increased in the CNS of EAE animals. These studies suggest that PGN and possibly other TLR agonists are capable of accessing the CNS during EAE, which could facilitate the reactivation of myelin-reactive T cells in the target tissue in EAE and MS. More recent studies demonstrated that infection of mice with *S. pneumoniae* was capable of increasing the severity of EAE, although this depended on the timing of infection relative to immunization (Herrmann et al. 2006). Furthermore, these studies demonstrated that the effects of *S. pneumoniae* on EAE were TLR2 dependent, as TLR2-deficient animals did not develop more severe EAE.

TLR4 can trigger both MyD88-dependent and MyD88-independent signaling, which may help to explain the somewhat contradictory findings concerning the effects of TLR4 agonists in modulating EAE. The Kuchroo laboratory utilized mice that express a transgenic TCR (5B6) on a B10.S background that do not develop EAE in spite of exhibiting a high frequency of autoreactive T cells. They demonstrated that these animals did not develop EAE due to the presence of APCs with limited T cell activating capacity (Waldner et al. 2004). These studies further demonstrated that activation of these APCs with TLR4 or TLR9 ligands broke T cell tolerance in these animals, resulting in the development of EAE. Studies by the Kubes laboratory indicated that pertussis toxin, which is commonly co-administered during immunization protocols for EAE, stimulates TLR4 signaling pathways. In addition, pertussis toxin also induced P-selectin expression, increased leukocyte/endothelial cell interactions, and facilitated T cell infiltration into the CNS. Pertussis toxin-mediated signaling and leukocyte extravasation into the CNS were found to be controlled by TLR4, as these effects were not observed in TLR4 knockout mice. The role of TLR4 in pertussis toxin induction of EAE was less clear, however. In general, TLR4 knockout mice were less susceptible to pertussis toxin-induced EAE than wild-type mice. However, this observation varied between individual experiments. In addition, TLR4 knockout mice that developed EAE in response to pertussis toxin developed disability scores that were almost as severe as wild-type animals. These studies suggest that pertussis toxin could be stimulating the TLR4 signaling of both the MyD88-dependent and the MyD88-independent pathways, and/or that pertussis toxin modulation of EAE is only partly dependent on TLR4 (Kerfoot et al. 2004; Racke et al. 2005). Studies in Lewis rats indicated that a combination of the TLR4 agonist LPS and the TLR9 agonist CpG ODN are required for the development of EAE, and that either agent alone added to IFA was capable of eliciting disease. Furthermore, a combination of CpG ODN and the TLR3 agonist poly I:C did not induce EAE (Wolf et al. 2007). These studies support the idea that stimulation of MyD88-independent TLR signaling suppresses EAE.

IRAK-1 plays a critical role in the MyD88-dependent signaling pathway. Interestingly, IRAK-1 knockout mice are resistant to the development of EAE (Deng et al. 2003). T cells derived from IRAK-1 knockout mice have normal TCR signaling but impaired Th1 cell development, suggesting that IRAK-1 is critical for proper T cell priming in the periphery. These conclusions are supported by subsequent studies indicating that although IRAK-1 mice are resistant to disease development using an active immunization protocol, Vα2.3/Vβ8.2 TCR transgenic T cells adoptively transfer disease in IRAK-1 knockout mouse recipients in a manner similar to wild-type recipients. Collectively, these findings suggest that IRAK-1 is critical for priming autoreactive T cells, but IRAK-1 expression is not required in the CNS for disease to occur (Hansen et al. 2006).

MyD88 knockout mice are resistant to the development of active EAE, further supporting a role of MyD88-dependent signaling in disease development (Prinz et al. 2006). Interestingly, T cells derived from MyD88 knockout mice did not respond measurably to their cognate antigen, suggesting that MyD88 knockout mice do not develop active EAE (at least in part) due to inadequate T cell priming in the periphery. However, bone marrow chimera studies indicated that MyD88 expression in the CNS also plays a significant role in controlling the development of EAE. Finally, adoptive transfer of myelin-specific T cells into MyD88 knockout mice resulted in decreased severity of EAE relative to wild-type animals, further supporting an important role for CNS expression of MyD88 in encephalitogenicity during the effector phase of EAE (Prinz et al. 2006).

Evidence suggests that TLR signaling through the MyD88-independent pathway either does not support or suppresses the development of EAE. For example, the TLR3 agonist polyinosinic:polycytidylic acid (poly I:C) does not support the development of active EAE when immunized with myelin antigens emulsified in IFA (Hansen et al. 2006). Furthermore, treatment of EAE mice with poly I:C suppressed the development of disease (Touil et al. 2006). In these studies, poly I:C treatment resulted in significant production of IFN-β, one of the critical products of the MyD88-independent signaling pathway. Significantly, IFN-β is commonly used in the treatment of MS. Recent studies further support a role of MyD88-independent signaling in suppression of EAE (Guo et al. 2008). In these studies, type I IFN receptor knockout mice developed more severe EAE than wild-type mice. Likewise, TRIF knockout mice lacking this critical adaptor molecule for MyD88-independent signaling also developed more severe disease. The studies further suggested that IFN-β-induced production of IL-27 by cells of the innate immune system played a critical role in suppressing the development of Th17 cells critical to disease development, and that this control was lost in type I IFN receptor and TRIF-deficient animals.

The expression of a variety of TLRs is elevated in the CNS of EAE mice (Zekki et al. 2002; Prinz et al. 2006; Xu and Drew 2007). Agents that specifically modulate TLR signaling pathways may be effective in the treatment of EAE and MS. Recently, we have evaluated the effects of peroxisome proliferator-activated receptors (PPARs), which are members of the nuclear hormone receptor family, on the expression of TLR intermediates by cells of the CNS. We demonstrated that PPAR-α agonists suppress the expression of critical MyD88-dependent signaling intermediates by primary microglia, as well as in the CNS of mice with EAE (Xu et al. 2007). PPAR-γ agonists also suppress the expression of MyD88 signaling intermediates by primary microglia (Xu and Drew 2007). In addition, these PPAR agonists suppressed glial production of IL-12 and IL-23, which are known to play critical roles in the development of Th1 and Th17 cells that stimulate the development of EAE. These studies suggest that PPAR agonists may be effective in the treatment of MS.

4 Summary

Environmental factors including pathogens are believed to stimulate the development of MS. It is clear that TLRs play a significant role in modulating disease. Studies to date suggest that TLRs, which activate MyD88-dependent signaling, contribute to the development of MS, whereas MyD88-independent pathways may mitigate disease severity. TLRs present on cells of the innate immune system are believed to provide critical signals involved in the activation of cells of the adaptive immune system, including autoreactive lymphocytes. However, many questions concerning the role of TLRs in modulating MS remain unanswered. The complex pattern of TLR expression in the periphery and in the CNS, as well as by cells of both the innate and adaptive immune systems, is just beginning to be appreciated. Future studies are needed to understand the detailed mechanisms by which TLRs modulate MS. Importantly, these studies could identify new targets for the treatment of MS.

Acknowledgments This work was supported by grants from the National Institutes of Health and the National Multiple Sclerosis Society.

References

Anderson DW, Ellenberg JH, Leventhal CM, Reingold SC, Rodriguez M, Silberberg DH (1992) Revised estimate of the prevalence of multiple sclerosis in the United States. Ann Neurol 31:333–336

Ando DG, Clayton J, Kono D, Urban JL, Sercarz EE (1989) Encephalitogenic T cells in the B10. PL model of experimental allergic encephalomyelitis (EAE) are of the Th-1 lymphokine subtype. Cell Immunol 124:132–143

Arnason BG (1983) Relevance of experimental allergic encephalomyelitis to multiple sclerosis. Neurol Clin 1:765–782

Bailey SL, Schreiner B, McMahon EJ, Miller SD (2007) CNS myeloid DCs presenting endogenous myelin peptides 'preferentially' polarize CD4 + T(H)-17 cells in relapsing EAE. Nat Immunol 8:172–180

Bell JK, Botos I, Hall PR, Askins J, Shiloach J, Segal DM, Davies DR (2005) The molecular structure of the Toll-like receptor 3 ligand-binding domain. Proc Natl Acad Sci USA 102:10976–10980

Bettelli E, Korn T, Oukka M, Kuchroo VK (2008) Induction and effector functions of T(H)17 cells. Nature 453:1051–1057

Brennan CA, Anderson KV (2004) Drosophila: the genetics of innate immune recognition and response. Annu Rev Immunol 22:457–483

Brown AM, McFarlin DE (1981) Relapsing experimental allergic encephalomyelitis in the SJL/J mouse. Lab Invest 45:278–284

Bsibsi M, Persoon-Deen C, Verwer RW, Meeuwsen S, Ravid R, Van Noort JM (2006) Toll-like receptor 3 on adult human astrocytes triggers production of neuroprotective mediators. Glia 53:688–695

Bsibsi M, Ravid R, Gveric D, van Noort JM (2002) Broad expression of Toll-like receptors in the human central nervous system. J Neuropathol Exp Neurol 61:1013–1021

Cao Z, Henzel WJ, Gao X (1996a) IRAK: a kinase associated with the interleukin-1 receptor. Science 271:1128–1131

Cao Z, Xiong J, Takeuchi M, Kurama T, Goeddel DV (1996b) TRAF6 is a signal transducer for interleukin-1. Nature 383:443–446
Compston A (2004) The pathogenesis and basis for treatment in multiple sclerosis. Clin Neurol Neurosurg 106:246–248
Compston DA, Vakarelis BN, Paul E, McDonald WI, Batchelor JR, Mims CA (1986) Viral infection in patients with multiple sclerosis and HLA-DR matched controls. Brain 109(Pt 2): 325–344
Deng C, Radu C, Diab A, Tsen MF, Hussain R, Cowdery JS, Racke MK, Thomas JA (2003) IL-1 receptor-associated kinase 1 regulates susceptibility to organ-specific autoimmunity. J Immunol 170:2833–2842
Du C, Yao SY, Ljunggren-Rose A, Sriram S (2002) *Chlamydia pneumoniae* infection of the central nervous system worsens experimental allergic encephalitis. J Exp Med 196:1639–1644
Fitzgerald KA, McWhirter SM, Faia KL, Rowe DC, Latz E, Golenbock DT, Coyle AJ, Liao SM, Maniatis T (2003a) IKKepsilon and TBK1 are essential components of the IRF3 signaling pathway. Nat Immunol 4:491–496
Fitzgerald KA, Rowe DC, Barnes BJ, Caffrey DR, Visintin A, Latz E, Monks B, Pitha PM, Golenbock DT (2003b) LPS-TLR4 signaling to IRF-3/7 and NF-kappaB involves the toll adapters TRAM and TRIF. J Exp Med 198:1043–1055
Frohman EM, Filippi M, Stuve O, Waxman SG, Corboy J, Phillips JT, Lucchinetti C, Wilken J, Karandikar N, Hemmer B, Monson N, De Keyser J, Hartung H, Steinman L, Oksenberg JR, Cree BA, Hauser S, Racke MK (2005) Characterizing the mechanisms of progression in multiple sclerosis: evidence and new hypotheses for future directions. Arch Neurol 62:1345–1356
Frohman EM, Racke MK, Raine CS (2006) Multiple sclerosis: the plaque and its pathogenesis. N Engl J Med 354:942–955
Gay D, Dick G, Upton G (1986) Multiple sclerosis associated with sinusitis: case-controlled study in general practice. Lancet 1:815–819
Goverman J, Woods A, Larson L, Weiner LP, Hood L, Zaller DM (1993) Transgenic mice that express a myelin basic protein-specific T cell receptor develop spontaneous autoimmunity. Cell 72:551–560
Guo B, Chang EY, Cheng G (2008) The type I IFN induction pathway constrains Th17-mediated autoimmune inflammation in mice. J Clin Invest 118:1680–1690
Hacker H, Redecke V, Blagoev B, Kratchmarova I, Hsu LC, Wang GG, Kamps MP, Raz E, Wagner H, Hacker G, Mann M, Karin M (2006) Specificity in Toll-like receptor signalling through distinct effector functions of TRAF3 and TRAF6. Nature 439:204–207
Hacker H, Vabulas RM, Takeuchi O, Hoshino K, Akira S, Wagner H (2000) Immune cell activation by bacterial CpG-DNA through myeloid differentiation marker 88 and tumor necrosis factor receptor-associated factor (TRAF)6. J Exp Med 192:595–600
Hansen BS, Hussain RZ, Lovett-Racke AE, Thomas JA, Racke MK (2006) Multiple toll-like receptor agonists act as potent adjuvants in the induction of autoimmunity. J Neuroimmunol 172:94–103
Hauser SL, Waubant E, Arnold DL, Vollmer T, Antel J, Fox RJ, Bar-Or A, Panzara M, Sarkar N, Agarwal S, Langer-Gould A, Smith CH (2008) B-cell depletion with rituximab in relapsing-remitting multiple sclerosis. N Engl J Med 358:676–688
Hausmann S, Wucherpfennig KW (1997) Activation of autoreactive T cells by peptides from human pathogens. Curr Opin Immunol 9:831–838
Hayashi F, Smith KD, Ozinsky A, Hawn TR, Yi EC, Goodlett DR, Eng JK, Akira S, Underhill DM, Aderem A (2001) The innate immune response to bacterial flagellin is mediated by Toll-like receptor 5. Nature 410:1099–1103
Hemmi H, Kaisho T, Takeuchi O, Sato S, Sanjo H, Hoshino K, Horiuchi T, Tomizawa H, Takeda K, Akira S (2002) Small anti-viral compounds activate immune cells via the TLR7 MyD88-dependent signaling pathway. Nat Immunol 3:196–200
Herrmann I, Kellert M, Schmidt H, Mildner A, Hanisch UK, Bruck W, Prinz M, Nau R (2006) *Streptococcus pneumoniae* Infection aggravates experimental autoimmune encephalomyelitis via Toll-like receptor 2. Infect Immun 74:4841–4848

Horwitz MS, Sarvetnick N (1999) Viruses, host responses, and autoimmunity. Immunol Rev 169:241–253

Jacobs LD, Cookfair DL, Rudick RA, Herndon RM, Richert JR, Salazar AM, Fischer JS, Goodkin DE, Granger CV, Simon JH, Alam JJ, Bartoszak DM, Bourdette DN, Braiman J, Brownscheidle CM, Coats ME, Cohan SL, Dougherty DS, Kinkel RP, Mass MK, Munschauer FE 3rd, Priore RL, Pullicino PM, Scherokman BJ, Whitham RH, et al. (1996) Intramuscular interferon beta-1a for disease progression in relapsing multiple sclerosis. The Multiple Sclerosis Collaborative Research Group (MSCRG). Ann Neurol 39:285–294

Johnson AJ, Suidan GL, McDole J, Pirko I (2007) The CD8 T cell in multiple sclerosis: suppressor cell or mediator of neuropathology? Int Rev Neurobiol 79:73–97

Johnson KP, Brooks BR, Cohen JA, Ford CC, Goldstein J, Lisak RP, Myers LW, Panitch HS, Rose JW, Schiffer RB (1995) Copolymer 1 reduces relapse rate and improves disability in relapsing-remitting multiple sclerosis: results of a phase III multicenter, double-blind placebo-controlled trial. The Copolymer 1 Multiple Sclerosis Study Group. Neurology 45:1268–1276

Kabelitz D (2007) Expression and function of Toll-like receptors in T lymphocytes. Curr Opin Immunol 19:39–45

Katz-Levy Y, Neville KL, Girvin AM, Vanderlugt CL, Pope JG, Tan LJ, Miller SD (1999) Endogenous presentation of self myelin epitopes by CNS-resident APCs in Theiler's virus-infected mice. J Clin Invest 104:599–610

Kawai T, Adachi O, Ogawa T, Takeda K, Akira S (1999) Unresponsiveness of MyD88-deficient mice to endotoxin. Immunity 11:115–122

Kawai T, Takeuchi O, Fujita T, Inoue J, Muhlradt PF, Sato S, Hoshino K, Akira S (2001) Lipopolysaccharide stimulates the MyD88-independent pathway and results in activation of IFN-regulatory factor 3 and the expression of a subset of lipopolysaccharide-inducible genes. J Immunol 167:5887–5894

Kennedy MK, Torrance DS, Picha KS, Mohler KM (1992) Analysis of cytokine mRNA expression in the central nervous system of mice with experimental autoimmune encephalomyelitis reveals that IL-10 mRNA expression correlates with recovery. J Immunol 149:2496–2505

Kerfoot SM, Long EM, Hickey MJ, Andonegui G, Lapointe BM, Zanardo RC, Bonder C, James WG, Robbins SM, Kubes P (2004) TLR4 contributes to disease-inducing mechanisms resulting in central nervous system autoimmune disease. J Immunol 173:7070–7077

Khoruts A, Miller SD, Jenkins MK (1995) Neuroantigen-specific Th2 cells are inefficient suppressors of experimental autoimmune encephalomyelitis induced by effector Th1 cells. J Immunol 155:5011–5017

Kurtzke JF (1993) Epidemiologic evidence for multiple sclerosis as an infection. Clin Microbiol Rev 6:382–427

Lampropoulou V, Hoehlig K, Roch T, Neves P, Calderon Gomez E, Sweenie CH, Hao Y, Freitas AA, Steinhoff U, Anderton SM, Fillatreau S (2008) TLR-activated B cells suppress T cell-mediated autoimmunity. J Immunol 180:4763–4773

Lehnardt S, Lachance C, Patrizi S, Lefebvre S, Follett PL, Jensen FE, Rosenberg PA, Volpe JJ, Vartanian T (2002) The toll-like receptor TLR4 is necessary for lipopolysaccharide-induced oligodendrocyte injury in the CNS. J Neurosci 22:2478–2486

Lehnardt S, Massillon L, Follett P, Jensen FE, Ratan R, Rosenberg PA, Volpe JJ, Vartanian T (2003) Activation of innate immunity in the CNS triggers neurodegeneration through a Toll-like receptor 4-dependent pathway. Proc Natl Acad Sci USA 100:8514–8519

Lenz DC, Lu L, Conant SB, Wolf NA, Gerard HC, Whittum-Hudson JA, Hudson AP, Swanborg RH (2001) A *Chlamydia pneumoniae*-specific peptide induces experimental autoimmune encephalomyelitis in rats. J Immunol 167:1803–1808

Li S, Strelow A, Fontana EJ, Wesche H (2002) IRAK-4: a novel member of the IRAK family with the properties of an IRAK-kinase. Proc Natl Acad Sci USA 99:5567–5572

Lublin FD (1982) Delayed, relapsing experimental allergic encephalomyelitis in mice. Role of adjuvants and pertussis vaccine. J Neurol Sci 57:105–110

Martin R, McFarland HF, McFarlin DE (1992) Immunological aspects of demyelinating diseases. Annu Rev Immunol 10:153–187

McFarland HF, Martin R (2007) Multiple sclerosis: a complicated picture of autoimmunity. Nat Immunol 8:913–919

McMahon EJ, Bailey SL, Castenada CV, Waldner H, Miller SD (2005) Epitope spreading initiates in the CNS in two mouse models of multiple sclerosis. Nat Med 11:335–339

Medzhitov R, Preston-Hurlburt P, Kopp E, Stadlen A, Chen C, Ghosh S, Janeway CA, Jr. (1998) MyD88 is an adaptor protein in the hToll/IL-1 receptor family signaling pathways. Mol Cell 2:253–258

Millefiorini E, Gasperini C, Pozzilli C, D'Andrea F, Bastianello S, Trojano M, Morino S, Morra VB, Bozzao A, Calo A, Bernini ML, Gambi D, Prencipe M (1997) Randomized placebo-controlled trial of mitoxantrone in relapsing-remitting multiple sclerosis: 24-month clinical and MRI outcome. J Neurol 244:153–159

Miller SD, Karpus WJ (1994) The immunopathogenesis and regulation of T-cell-mediated demyelinating diseases. Immunol Today 15:356–361

Miller SD, Vanderlugt CL, Begolka WS, Pao W, Yauch RL, Neville KL, Katz-Levy Y, Carrizosa A, Kim BS (1997) Persistent infection with Theiler's virus leads to CNS autoimmunity via epitope spreading. Nat Med 3:1133–1136

Mosmann TR, Coffman RL (1989) TH1 and TH2 cells: different patterns of lymphokine secretion lead to different functional properties. Annu Rev Immunol 7:145–173

Muzio M, Natoli G, Saccani S, Levrero M, Mantovani A (1998) The human toll signaling pathway: divergence of nuclear factor kappaB and JNK/SAPK activation upstream of tumor necrosis factor receptor-associated factor 6 (TRAF6). J Exp Med 187:2097–2101

Nair A, Frederick TJ, Miller SD (2008) Astrocytes in multiple sclerosis: a product of their environment. Cell Mol Life Sci 65:2702–2720

Nikbin B, Bonab MM, Khosravi F, Talebian F (2007) Role of B cells in pathogenesis of multiple sclerosis. Int Rev Neurobiol 79:13–42

Noseworthy JH (1999) Progress in determining the causes and treatment of multiple sclerosis. Nature 399:A40–A47

Oldstone MB (1987) Molecular mimicry and autoimmune disease. Cell 50:819–820

Oshiumi H, Matsumoto M, Funami K, Akazawa T, Seya T (2003) TICAM-1, an adaptor molecule that participates in Toll-like receptor 3-mediated interferon-beta induction. Nat Immunol 4:161–167

Panitch HS (1994) Influence of infection on exacerbations of multiple sclerosis. Ann Neurol 36(Suppl):S25–S28

Prinz M, Garbe F, Schmidt H, Mildner A, Gutcher I, Wolter K, Piesche M, Schroers R, Weiss E, Kirschning CJ, Rochford CD, Bruck W, Becher B (2006) Innate immunity mediated by TLR9 modulates pathogenicity in an animal model of multiple sclerosis. J Clin Invest 116:456–464

Racke MK (2001) Experimental autoimmune encephalomyelitis (EAE). Curr Protoc Neurosci Chap 9:Unit 9.7

Racke MK, Hu W, Lovett-Racke AE (2005) PTX cruiser: driving autoimmunity via TLR4. Trends Immunol 26:289–291

Rapp NS, Gilroy J, Lerner AM (1995) Role of bacterial infection in exacerbation of multiple sclerosis. Am J Phys Med Rehabil 74:415–418

Rocken M, Racke M, Shevach EM (1996) IL-4-induced immune deviation as antigen-specific therapy for inflammatory autoimmune disease. Immunol Today 17:225–231

Rowe DC, McGettrick AF, Latz E, Monks BG, Gay NJ, Yamamoto M, Akira S, O'Neill LA, Fitzgerald KA, Golenbock DT (2006) The myristoylation of TRIF-related adaptor molecule is essential for Toll-like receptor 4 signal transduction. Proc Natl Acad Sci USA 103:6299–6304

Sadovnick AD, Ebers GC (1993) Epidemiology of multiple sclerosis: a critical overview. Can J Neurol Sci 20:17–29

Sanders P, De Keyser J (2007) Janus faces of microglia in multiple sclerosis. Brain Res Rev 54:274–285

Schnare M, Holt AC, Takeda K, Akira S, Medzhitov R (2000) Recognition of CpG DNA is mediated by signaling pathways dependent on the adaptor protein MyD88. Curr Biol 10:1139–1142

Segal BM, Chang JT, Shevach EM (2000) CpG oligonucleotides are potent adjuvants for the activation of autoreactive encephalitogenic T cells in vivo. J Immunol 164:5683–5688

Sibley WA, Bamford CR, Clark K (1985) Clinical viral infections and multiple sclerosis. Lancet 1:1313–1315

Sriram S, Mitchell W, Stratton C (1998) Multiple sclerosis associated with *Chlamydia pneumoniae* infection of the CNS. Neurology 50:571–572

Takeda K, Akira S (2005) Toll-like receptors in innate immunity. Int Immunol 17:1–14

Takeda K, Kaisho T, Akira S (2003) Toll-like receptors. Annu Rev Immunol 21:335–376

Takeuchi O, Kaufmann A, Grote K, Kawai T, Hoshino K, Morr M, Muhlradt PF, Akira S (2000) Cutting edge: preferentially the *R*-stereoisomer of the mycoplasmal lipopeptide macrophage-activating lipopeptide-2 activates immune cells through a toll-like receptor 2- and MyD88-dependent signaling pathway. J Immunol 164:554–557

Tauber SC, Nau R, Gerber J (2007) Systemic infections in multiple sclerosis and experimental autoimmune encephalomyelitis. Arch Physiol Biochem 113:124–130

Touil T, Fitzgerald D, Zhang GX, Rostami A, Gran B (2006) Cutting edge: TLR3 stimulation suppresses experimental autoimmune encephalomyelitis by inducing endogenous IFN-beta. J Immunol 177:7505–7509

Vanderlugt CL, Miller SD (2002) Epitope spreading in immune-mediated diseases: implications for immunotherapy. Nat Rev Immunol 2:85–95

Visser L, Jan de Heer H, Boven LA, van Riel D, van Meurs M, Melief MJ, Zahringer U, van Strijp J, Lambrecht BN, Nieuwenhuis EE, Laman JD (2005) Proinflammatory bacterial peptidoglycan as a cofactor for the development of central nervous system autoimmune disease. J Immunol 174:808–816

Visser L, Melief MJ, van Riel D, van Meurs M, Sick EA, Inamura S, Bajramovic JJ, Amor S, Hintzen RQ, Boven LA, t Hart BA, Laman JD (2006) Phagocytes containing a disease-promoting Toll-like receptor/Nod ligand are present in the brain during demyelinating disease in primates. Am J Pathol 169:1671–1685

Waldner H, Collins M, Kuchroo VK (2004) Activation of antigen-presenting cells by microbial products breaks self tolerance and induces autoimmune disease. J Clin Invest 113:990–997

Wolf NA, Amouzegar TK, Swanborg RH (2007) Synergistic interaction between Toll-like receptor agonists is required for induction of experimental autoimmune encephalomyelitis in Lewis rats. J Neuroimmunol 185:115–122

Xu J, Drew PD (2007) Peroxisome proliferator-activated receptor-gamma agonists suppress the production of IL-12 family cytokines by activated glia. J Immunol 178:1904–1913

Xu J, Racke MK, Drew PD (2007) Peroxisome proliferator-activated receptor-alpha agonist fenofibrate regulates IL-12 family cytokine expression in the CNS: relevance to multiple sclerosis. J Neurochem 103:1801–1810

Yamamoto M, Sato S, Hemmi H, Hoshino K, Kaisho T, Sanjo H, Takeuchi O, Sugiyama M, Okabe M, Takeda K, Akira S (2003) Role of adaptor TRIF in the MyD88-independent toll-like receptor signaling pathway. Science 301:640–643

Yamin TT, Miller DK (1997) The interleukin-1 receptor-associated kinase is degraded by proteasomes following its phosphorylation. J Biol Chem 272:21540–21547

Zekki H, Feinstein DL, Rivest S (2002) The clinical course of experimental autoimmune encephalomyelitis is associated with a profound and sustained transcriptional activation of the genes encoding toll-like receptor 2 and CD14 in the mouse CNS. Brain Pathol 12:308–319

Toll-Like Receptors in Peripheral Nerve Injury and Neuropathic Pain*

Donghoon Kim, Soojin Lee, and Sung Joong Lee

Contents

Abstract Peripheral nerve injury triggers a series of responses in the injured nerve, such as the dissolution of distal axons, the activation of Schwann cells, the production of various proinflammatory mediators, and the infiltration of circulating immune cells. These orchestrated events regulate the degeneration and subsequent regeneration of the injured nerve. In addition, peripheral nerve injury often accompanies chronic pain. Studies in this field have revealed that spinal cord microglia activation plays a critical role in the development of pain hypersensitivity. Recent studies using genetically modified mice indicate that Toll-like receptors (TLRs) are involved in nerve degeneration (Wallerian degeneration) and chronic pain (neuropathic pain) development after nerve injury. Here, we review studies that

D. Kim and S.J. Lee (✉)
Program in Molecular and Cellular Neuroscience, DRI, BK21, and Department of Oral Physiology, School of Dentistry, Seoul National University, Seoul, Republic of Korea
e-mail: sjlee87@snu.ac.kr

S. Lee
Department of Microbiology, School of Systems Biology, Chungnam National University, Daejoen, Republic of Korea

*D. Kim and S. Lee contributed equally in writing this chapter

T. Kielian (ed.), *Toll-like Receptors: Roles in Infection and Neuropathology*, Current Topics in Microbiology and Immunology 336,
DOI 10.1007/978-3-642-00549-7_10, © Springer-Verlag Berlin Heidelberg 2009

have implicated TLRs in mediating nerve degeneration/regeneration and neuropathic pain following nerve injury. In addition, we discuss possible mechanisms underlying the roles of TLRs in these neurological disorders.

Abbreviations

ATP	Adenosine triphosphate
BDNF	Brain-derived neurotrophic factor
CNTF	Ciliary neurotrophic factor
COX-2	Cyclooxygenase 2
DRG	Dorsal root ganglia
FGF	Fibroblast growth factor
GDNF	Glial cell line-derived neurotrophic factor
HMGB1	High mobility group box 1 protein
HSP	Heat-shock protein
iNOS	Inducible nitric oxide synthase
LIF	Leukemia inhibitory factor
LPA	Lysophosphatidic acid
MCP-1	Monocyte chemoattractant protein 1 (CCL2)
MIP-1α	Macrophage inflammatory protein 1α (CCL3)
NGF	Nerve growth factor
NO	Nitric oxide
Poly I:C	Polyinosinic:polycytidylic acid
PNS	Peripheral nervous system
RANTES	Regulated upon activation, normal T cell expressed and secreted (CCL5)
TLR	Toll-like receptor

1 Introduction

The peripheral nerves of our body can become damaged for various reasons, including traumatic nerve injury due to accidents, viral or bacterial infection, or metabolic disorders such as diabetes mellitus (Tsuda et al. 2005). Such peripheral nerve injuries initiate an orchestrated sequence of pathobiological events in the injured nerve that eventually result in the degeneration/regeneration of damaged axons (Coleman 2005). In certain circumstances, these pathobiological events are followed by abnormal pain sensations that are collectively called "neuropathic pain" (Merskey 1994). Many studies have shown that these nerve degeneration and regeneration processes are accompanied by local inflammatory responses at the injury site and the dorsal root ganglia (DRG) (Marchand et al. 2005). Blood-derived immune cells, nerve-resident macrophages, and Schwann cells all play important roles in this process. In addition, studies on neuropathic pain over the last decade have demonstrated that peripheral nerve injury also induces a local inflammatory response in the spinal cord, which plays an important role in the development of abnormal pain sensations (Watkins et al. 2001).

Toll-like receptors (TLRs) are type I transmembrane glycoproteins expressed in the cells of the innate immune system that function as receptors for pathogen-derived molecules (Medzhitov 2001). Upon binding to their respective ligands, TLRs transmit intracellular signals that induce the expression of inflammatory genes, eventually leading to an innate immune response. Interestingly, studies on the function of TLRs showed that TLRs also recognize endogenous molecules that are released from damaged tissue. Therefore, in this context, it is more likely that TLRs sense "danger signals" rather than identify "non-self" molecules (Matzinger 2002). This suggests that TLRs may play roles in inflammatory responses in the injured peripheral nerve. Indeed, data are emerging that demonstrate a pivotal role for TLRs in peripheral nerve injury and neuropathic pain.

2 TLRs in Peripheral Nerve Injury

2.1 Wallerian Degeneration After Nerve Injury

Traumatic nerve injury, as well as other peripheral neuropathy, can cause nerve degeneration, which is typically referred to as "Wallerian degeneration" after the scientist who first described it. Studies over several decades have revealed that Wallerian degeneration is not merely the passive demise process of damaged neurons, but rather it is an active and orchestrated process requiring energy and a proteosome-dependent proteolytic pathway (Mack et al. 2001). Wallerian degeneration not only leads to the degeneration of injured axons, but it is also a prerequisite for the successful regeneration of the degenerated axons (Luk et al. 2003). Immediately after nerve injury, axons proximal to the cell body rapidly retract and transmit retrograde survival/stress signals to the cell body. It has been reported that a number of signaling molecules, such as importin, vimentin, and JNK are locally synthesized or activated at the injured axon terminal and are retrogradely transported to the cell body, where they activate stress-induced transcription factors such as AP-1 (Hanz et al. 2003; Lindwall and Kanje 2005). Such retrograde signals play important roles in the survival of the damaged neurons and regrowth of injured axons because inhibition of the retrograde signal enhances neuronal apoptosis due to nerve axotomy (Hanz et al. 2003; Schweizer et al. 2002).

There is also an orchestrated sequence of histopathological events that occur in the distal part of the damaged axons. In the distal axons, Schwann cells first respond to the nerve injury and become de-differentiated, or activated (Stoll et al. 2002). Upon activation, the expression of specific genes for myelin formation such as MBP, P0, and PMP22 is rapidly downregulated, while other sets of genes involved in cell proliferation and stress response are upregulated (Kuhn et al. 1993).

The activated Schwann cells then acquire a phagocytic capacity and begin to remove myelin and degenerating axons (Stoll et al. 2002). Such a cleaning process is essential for successful nerve regeneration, because the myelin debris contains molecules that inhibit the outgrowth of regenerating axons (Mukhopadhyay et al. 1994). In addition, activated Schwann cells express various inflammatory mediators, including cytokines (TNF-α, IL-6, and LIF), chemokines (MCP-1 and MIP-1α) and other enzymes (matrix metalloproteases, Cox-2, and iNOS) (Bolin et al. 1995; Chattopadhyay et al. 2007; Levy et al. 1999; Takahashi et al. 2004; Taskinen and Roytta 2000; Tofaris et al. 2002; Wagner and Myers 1996). Matrix metalloproteases produced by activated Schwann cells along with other vasoactive mediators such as nitric oxide (NO) may lead to an interruption of the blood–nerve barrier and invasion of circulating immune cells at the injury site (Shubayev et al. 2006; Zochodne et al. 1999). In addition, MCP-1 and LIF attract monocytes/macrophages to the injury site from the blood vessels and induce a local inflammatory response (Tofaris et al. 2002). These peripheral immune cells, in turn, remove the myelin debris and degenerating axons and complete the cleaning process that was initiated by the activated Schwann cells. In addition, Schwann cells and nerve-recruited macrophages produce various growth factors such as NGF, CNTF, GDNF, FGF and BDNF (Hammarberg et al. 1996; Meyer et al. 1992; Smith et al. 1993). GDNF and BDNF bind to their receptors on the axon terminal and are transported to the cell body, where they regulate the survival and axon growth of the damaged neurons (Boyd and Gordon 2003; Rind et al. 2005). Meanwhile, these growth factors as well as other inflammatory mediators, including cytokines, chemokines, NO and prostaglandins, may directly sensitize sensory neurons and thereby contribute to the development of nerve injury-induced pain (Cunha et al. 2005; Malin et al. 2006). At the same time, FGF induces Schwann cell proliferation underneath the basement membrane of the endoneurium (Jungnickel et al. 2006) and forms a band-like structure called the "band of Büngner." Along this structure, surviving neurons extend their regenerating axons and find their target for reinnervation. Finally, the dedifferentiated Schwann cells, upon contact with the regrowing axons, remyelinate axons and restore the physiological functions of reinnervated axons.

Therefore, it is clear that Schwann cell activation plays a critical role in Wallerian degeneration and successful nerve regeneration, but how a nerve injury initiates the activation of Schwann cells remains elusive. It was proposed that erbB2 activation on Schwann cells in direct contact with axons at nodal regions is a key event leading to Schwann cell activation after nerve injury (Guertin et al. 2005). In this study, a chemical ErbB2 antagonist inhibited demyelination of an injured nerve in vivo. In addition, activation of ErbB2 by treatment with neuregulin triggered demyelination of the axons in a DRG neuron–Schwann cell co-culture system. Still, there is no evidence of how nerve injury triggers ErbB2 activation in Schwann cells. In addition, evidences are lacking that ErbB2 activation induces TNF-α or IL-6 gene expression that is expressed by Schwann cells during Wallerian degeneration. This implies that additional signaling pathway(s) are employed to induce such inflammatory gene expression from Schwann cells. Therefore, it still remains to be elucidated how Schwann cells initially sense nerve injury to trigger the subsequent

Wallerian degeneration process. Recent studies on the role of TLRs in Schwann cell activation will shed new light on this issue (Karanth et al. 2006; Lee et al. 2006; Oliveira et al. 2003).

2.2 TLR Expression in the Peripheral Nervous System

TLRs were originally identified as receptors that recognize molecules derived from microorganisms. Therefore, studies have largely focused on the roles of TLRs in innate immune cells such as macrophages and dendritic cells (Medzhitov 2001), and little attention has been paid to their role in Schwann cells and neurons, the two major cell types constituting peripheral nerves. However, recent studies have shown that TLRs are also widely expressed in these cells, suggesting a putative TLR function in the peripheral nervous system (PNS).

TLR expression in Schwann cells was expected given that Schwann cells respond to LPS (Skundric et al. 1997) and TLR4 is a signaling receptor for LPS (Poltorak et al. 1998). However, direct evidence of TLR expression in Schwann cells had not been reported until the study by Oliveira et al. that demonstrated expression of TLR2 in primary cultured human Schwann cells and in Schwann cells from skin lesions of a leprosy patient (Oliveira et al. 2003). Our group also characterized the expression of TLRs in rat Schwann cells, and found that not only TLRs 2 and 4 but also TLR3 are highly expressed in primary rat Schwann cells at the mRNA and protein levels (Lee et al. 2007). Thus far, expression of other TLRs has not been reported in Schwann cells.

The initial studies investigating TLR expression in the nervous system indicated that TLRs are expressed in glial cells but not in neurons in the brain (Lehnardt et al. 2002). However, in more recent studies, it was found that TLRs are also expressed in different types of neurons. Neuronal TLR expression was first documented in a study by Prehaud et al., which demonstrated neuronal TLR expression by showing TLR3 expression in the human neuronal cell line, NT2-N (Prehaud et al. 2005). Virus infection or polyinosinic:polycytidylic acid (poly I:C) treatment induced typical antiviral gene expression in this cell line (Lafon et al. 2006), suggesting a role for neuronal TLR3 in virus-induced innate immune responses. Neuronal TLR3 expression was also confirmed in primary cultured mouse cortical and DRG neurons (Cameron et al. 2007) and human histopathologic tissue sections of a post-mortem brain (Jackson et al. 2006). In addition, TLRs 2 and 4 were induced after ischemic brain injury in mouse cortical neurons in vitro as well as in vivo (Tang et al. 2007). In this study, TLR expression in damaged cortical neurons contributed to apoptotic neuronal cell death. In another study, TLR8 was found to be expressed in various regions of the nervous system during mouse development, and TLR8 stimulation inhibited neurite outgrowth and induced neuronal apoptosis (Ma et al. 2006). Based on these findings, it appears that TLRs are differentially expressed in the nervous system depending on neuronal cell types and developmental stages. Among the TLRs detected in neurons, TLRs 3, 4, and 8 have been reported to be expressed in neurons of the PNS, such as DRG neurons and trigeminal sensory neurons (Cameron et al. 2007; Ma et al. 2006; Wadachi and Hargreaves 2006).

Macrophages are readily detected in the DRG and nerve tissues, and macrophage numbers increase greatly after nerve injury (Hu and McLachlan 2003). Considering the TLR expression profile of macrophages (Applequist et al. 2002), it is reasonable to speculate that PNS-resident macrophages express many different TLR members. Thus far, TLR expression in nerve-resident macrophages has not been reported.

2.3 TLRs in Traumatic Peripheral Nerve Injury

The recent finding that TLRs are expressed on Schwann cells in the PNS, combined with the fact that TLRs play a role in the detection of necrotic cells by innate immune cells (Li et al. 2001), suggests a putative role for TLRs in Schwann cell activation observed after nerve injury. We first explored this possibility by stimulating primary Schwann cells with necrotic neurons. We reasoned that damaged neurons would release molecules that might serve as endogenous TLR agonists. As speculated, stimulation of Schwann cells with necrotic neurons strongly induced the expression of various proinflammatory genes such as those for TNF-α, LIF, and MCP-1 (Lee et al. 2006), which are expressed in axotomized nerve in vivo after injury (Carroll and Frohnert 1998; Cui et al. 2000; Dowsing et al. 2001). It has been previously documented that LIF and MCP-1 expressed in activated Schwann cells can recruit macrophages to the injury site, a key feature of Wallerian degeneration (Tofaris et al. 2002), and TNF-α is implicated in the development of nerve injury-induced demyelination and pain hypersensitivity (Lindenlaub et al. 2000; Stoll et al. 1993). Such findings suggest that Schwann cell activation induced by necrotic neurons may serve as a mechanism underlying the activation of these cells following nerve injury. In addition, necrotic neuron-induced proinflammatory gene expression is completely blocked in Schwann cells of TLR2 knockout mice, and is also partially attenuated in cells from TLR3 knockout mice (Lee et al. 2006). These data indicate that TLRs 2 and 3 are indeed involved in the Schwann cell activation process, at least in vitro. Similarly, TLR4 was implicated in Schwann cell activation after nerve injury (Karanth et al. 2006). In this study, primary Schwann cells were activated by nerve homogenate, which was partially attenuated by a TLR4 blocking antibody. The findings of these in vitro reports were later supported by an in vivo study using TLR2- and TLR4-deficient mice (Boivin et al. 2007). Following sciatic nerve lesion, delayed Wallerian degeneration, axonal regeneration, and reduced locomotor recovery were observed in TLR2- and TLR4-deficient mice compared to wild-type animals. In addition, activation of TLRs 2 and 4 by injecting their respective agonists at the sciatic nerve accelerated myelin debris clearance and recovery of peripheral nerve function. Although these effects of TLR signaling cannot be directly attributed to TLR expression in Schwann cells, these in vivo data, in concert with the previous in vitro data of ours and from Karanth et al., strongly argue that TLRs expressed on Schwann cells may function as sensor molecules for nerve injury, which may initiate Schwann cell activation and subsequent Wallerian degeneration.

After nerve axotomy, the proximal axons of injured neurons retract rapidly to a certain distance from the injury site (Kerschensteiner et al. 2005). According to a study by Cameron et al., TLR3 is expressed in DRG neurons and, furthermore, its expression is concentrated in growth cones (Cameron et al. 2007). In addition, the activation of TLR3 in DRG neurons by poly I:C and mouse brain mRNA resulted in profound growth cone collapse (Cameron et al. 2007). These data suggest that TLR3-induced growth cone collapse may account for the axon retraction observed in the injured nerve.

2.4 TLRs in Infectious Peripheral Neuropathy

Peripheral nerves are also vulnerable to damage from infectious pathogens. For example, *Mycobacterium leprae* infection in the PNS causes nerve damage that is a hallmark of leprosy (Oliveira et al. 2003). In addition, virus infection in the PNS causes peripheral neuropathy that accompanies inflammation and neuronal cell death. For instance, human immunodeficiency virus (HIV) infection causes inflammation in the peripheral nerve, resulting in DRG neuronal cell death and severe neuropathic pain (Pardo et al. 2001). Similarly, herpes simplex virus (HSV) infection causes neuropathy in the facial and optic nerves (Yura 2000). Considering the involvement of TLRs as receptors in pathogen infection, it is conceivable that TLRs expressed on Schwann cells or neurons may also be involved in pathogen recognition in the PNS. Thus far, circumstantial evidence supports such a role of TLRs in infectious neuropathy. First, activation of TLR2 by a lipoprotein of *M. leprae* induces apoptosis of primary cultured human Schwann cells (Oliveira et al. 2003). Such Schwann cell responses to a TLR2 agonist may serve as one mechanism of Schwann cell apoptosis observed in leprosy in vivo.

In human viral encephalitis, TLR3 expression is upregulated in cerebellar cortical neurons in postmortem human brains (Jackson et al. 2006). TLR3 activation in a human neuronal cell line by rabies virus infection or poly I:C treatment results in an innate immune response culminating in the production of IFN-β, chemokines, and cytokines (Lafon et al. 2006; Prehaud et al. 2005). Furthermore, TLR3 knockout mice show enhanced mortality and increased viral replication in their brains after West Nile virus infection, suggesting an antiviral function of TLR3 in this model (Daffis et al. 2008). Although such a TLR3-medited anti-viral response was not reported in DRG neurons, these data suggest that neuronal TLR3 may initiate antiviral innate immune responses upon virus infection in the PNS. Similarly, TLR3 activation in Schwann cells may function in viral neuropathy. To test this hypothesis, we investigated the cellular response of TLR3 stimulation in primary rat Schwann cells. In our study, TLR3 stimulation induced the expression of various proinflammatory genes, including those for TNF-α, RANES, and IP-10 (unpublished data). In addition, TLR3 stimulation induced iNOS gene expression and NO production in primary rat Schwann cells (Lee et al. 2007). In a DRG neuron–Schwann cell co-culture, TLR3 stimulation induced DRG neuronal cell death, which was dependent on NO production from the activated Schwann cells. In addition, administration of a TLR agonist into rat sciatic nerves induced axonal demyelination in vivo (Lee et al. 2007). Such effects of poly I:C are reminiscent of DRG neuronal demise and demyelination observed during viral infection. These data argue that TLR3 expressed on Schwann cells may recognize double-stranded RNA derived from virus infection and initiate an inflammatory response in the PNS. Still, direct in vivo evidence of a role for TLRs in these infectious peripheral neuropathies is lacking. Future studies using TLR knockout mice, and preferably Schwann cell- or neuron-specific TLR conditional knockout mice, in infectious nerve injury models would conclusively address the in vivo role of Schwann cell or neuronal TLRs in infectious peripheral neuropathy.

3 TLRs in Neuropathic Pain

3.1 What is Neuropathic Pain?

Injury in the peripheral nerve often results in abnormal chronic pain sensations. Such pathological pain is generally referred to as "neuropathic pain," as opposed to inflammatory pain that is caused by tissue inflammation (Marchand et al. 2005). According to the definition offered by the International Association for the Study of Pain, neuropathic pain is "pain caused by a primary lesion or dysfunction in the nervous system" (Merskey 1994). The clinical symptoms of neuropathic pain include pain induction by innocuous stimuli that normally do not elicit pain, which is called allodynia; a greater and more prolonged pain sensation induced by noxious stimuli, which is referred to as hyperalgesia; or a spontaneous burning sensation without any stimuli (Marchand et al. 2005). Neuropathic pain typically develops only when sensory nerve fibers are affected, whereas injury in the motor fibers is generally painless (Koltzenburg 2005). It is known that such pathological pain conditions are also induced by viral infection, e.g., postherpetic neuralgia, and by other metabolic diseases such as diabetes (Tsuda et al. 2005).

In an effort to investigate the mechanisms of neuropathic pain, several animal models have been developed in rodents by chronic constriction injury (CCI) of the sciatic nerve (Bennett and Xie 1988), partial sciatic nerve ligation (Seltzer et al. 1990), spinal nerve ligation (Kim and Chung 1992), and lesions in the tibial and peroneal nerve sparing the sural nerve (Decosterd and Woolf 2000). In all these animal models, the peripheral nerve is subjected to partial injury, but it still retains its innervation to the peripheral tissue, which is required for stimulus-induced pain processing.

Upon partial peripheral nerve injuries, some or all of the above symptoms observed in neuropathic pain patients, such as allodynia and hyperalgesia, can be experimentally induced in these animal models, and the associated pain behaviors can be assessed by specifically designed pain test systems that include the hot plate and pin-prick tests (Malmberg and Bannon 1999).

3.2 Spinal Cord Glia as Key Players in Neuropathic Pain

Because pain signals, just like any other sensory modality, are transmitted to the brain via neurons, and neuropathic pain is generally elicited by neuronal damage, researchers have taken it for granted that neuropathic pain is caused mainly by the alteration of neurons of the pain-transmitting circuit. Thus, researchers have focused on elucidating the nature of injury-induced changes in the DRG and spinal cord pain-transmitting neurons. Based on these studies, several mechanisms have been proposed to explain nerve injury-induced pain hypersensitivity, including nerve injury-induced changes in the expression of channels and receptors in the DRG or spinal cord neurons (Dib-Hajj et al. 1999; Yang et al. 2004; Salter and Woolf 2005), aberrant sprouting of sympathetic axons in the DRG (Ramer and Bisby 1997), and the demise of inhibitory interneurons in the spinal cord (Sugimoto et al. 1990). However, there has been little focus on the contributions of glial cells, and recent discoveries about the roles of spinal cord glial cells have dramatically changed the traditional view on the mechanism of neuropathic pain.

The finding that peripheral nerve injury may induce astrocyte activation in the anatomically remote spinal cord was first reported by Gilmore and Skinner (1979), and spinal cord microglia activation was reported later. Since then, a series of studies have documented spinal cord glial cell activation in the context of various neuropathic pain models, including the spinal nerve transection, peripheral nerve trauma, and lumbar root constriction models (Colburn et al. 1999; Hashizume et al. 2000). In an effort to elucidate the role of glial cells in neuropathic pain, microglia were directly activated by an intrathecal injection of HIV coat protein (gp120) in the absence of peripheral nerve injury (Milligan et al. 2001). In this study, microglia activation alone was able to induce mechanical allodynia and hyperalgesia, key features of neuropathic pain. In addition, inhibition of glial cell activation with chemical inhibitors revealed that microglia activation is required for the induction of neuropathic pain (Ledeboer et al. 2005), whereas astrocyte activation is involved in the maintenance of pain hypersensitivity after nerve injury (Raghavendra et al. 2003). Based on these findings, it is now generally accepted that glial cell activation after peripheral nerve injury plays a critical role in the initiation and/or maintenance of neuropathic pain (Watkins et al. 2001). Thus far, it is not completely known how spinal cord glial cell activation contributes to the initiation and maintenance of neuropathic pain. Several lines of evidence indicate that proinflammatory mediators expressed by activated glial cells are involved in these processes. Upon peripheral nerve injury, various proinflammatory mediators such as TNF-α, IL-1β, IL-6, iNOS, and COX-2 are upregulated in activated spinal cord microglia and/or astrocytes (Broom et al. 2004; DeLeo et al. 1996; Hashizume et al. 2000; Levy and Zochodne 1998; Ohtori et al. 2004). Blocking the expression and/or function of these proinflammatory molecules by the injection of either neutralizing antibodies, soluble receptors, or chemical inhibitors can attenuate nerve injury-induced pain hypersensitivity in vivo (Arruda et al. 2000; Ma et al. 2002; Meller et al. 1992; Sommer et al. 2001; Sweitzer et al. 2001). It is conceivable that these inflammatory mediators released from activated glia in the spinal cord may directly affect the excitability of the pain-transmitting neurons by regulating gene expression (McMahon et al. 2005). It was also reported that BDNF released from activated microglia produces a depolarizing shift in the anion reversal potential in spinal lamina I neurons. This shift inverts the polarity of currents activated by GABA, which may diminish the inhibitory input to the pain-transmitting neurons and thereby contribute to pain transmission (Coull et al. 2005). It is also possible that inflammatory mediators lead to the loss of inhibitory interneuronal function in the spinal cord (Vikman et al. 2007). This would prevent the inhibition of pain-transmitting neurons, thereby inducing pain hypersensitivity (Salter and Woolf 2005). In either case, previous studies imply that proinflammatory mediators released from the activated spinal cord glia induce hypersensitivity in pain-transmitting neurons.

3.3 Involvement of TLRs in Neuropathic Pain

While it is generally accepted that spinal cord microglia activation leads to neuropathic pain, there is still debate as to how spinal cord glial cells are activated upon peripheral nerve injury, thereby causing pain hypersensitivity. Recent studies on the role of TLRs in nerve injury-induced spinal cord microglia activation shed new light on this issue. The evidence for a role of TLRs in neuropathic pain was first reported in a study by Tanga et al., which demonstrated TLR4 mRNA upregulation in spinal cord microglia upon L5 nerve transection (Tanga et al. 2004). Later, this same group demonstrated that TLR4 expression is required for nerve injury-induced spinal cord glia activation and tactile and thermal hypersensitivity (Tanga et al. 2005). By intrathecal injection of antisense TLR4 oligodeoxynucleotides, these authors demonstrated that the pain attenuation observed in TLR4 knockout mice is due to TLR4 inhibition in spinal cord cells, most probably in microglia. These data suggest that nerve injury may trigger microglia activation via TLR4. More recently, it was reported that TLR4 is important not only for initiation but also for the maintenance of spinal cord microglia activation and neuropathic pain (Hutchinson et al. 2008).

After the initial discovery of the pivotal role of TLR4 in neuropathic pain, the involvement of other TLR members (TLRs 2 and 3) was investigated. Similar to TLR4 knockout mice, nerve injury-induced mechanical allodynia and thermal hyperalgesia are reduced in TLR2 knockout mice compared to wild-type animals (Kim et al. 2007). Likewise, the nerve injury-induced spinal cord microglia and astrocyte activation that accompany the induction of pain-related gene expression are reduced in TLR2 knockout mice. Interestingly, in TLR3 knockout mice and TLR3 antisense oligodeoxynucleotide-injected mice, mechanical allodynia, but not thermal hyperalgesia, was attenuated (Obata et al. 2008). This suggests that the mechanisms underlying the effects of TLR3 on neuropathic pain may be distinct from those for TLRs 2 and 4. In this regard, it is interesting to note that knockout of TLR3 fails to inhibit nerve injury-induced astrocyte activation, whereas knockout of TLR2 or TLR4 suppresses both astrocyte and microglia activation. Studies so far indicate that microglia constitutively express TLRs 2, 3 and 4 in vitro as well as in vivo (Babcock et al. 2006; Chakravarty and Herkenham 2005; Town et al. 2006). All of these TLR members can function as receptors for damaged neurons (Karanth et al. 2006; Lee et al. 2006). This may account for the partial recovery from pain hypersensitivity observed in TLR2, TLR3, and TLR4 knockout mice. It will be interesting to test if nerve injury-induced thermal hyperalgesia is completely abrogated in TLR2/4 double-knockout mice, and if mechanical allodynia is completed blocked in TLR2/3/4 triple-knockout mice.

Based on our studies and those of other researchers, it can be postulated that peripheral nerve injury results in the release of endogenous TLR agonists in the spinal cord. These agonists activate spinal cord microglia via TLRs that lead to the enhanced expression of proinflammatory mediators, which, in turn, affects neurons, thereby causing pain hypersensitivity. However, these data do not exclude the possibility that TLRs expressed on the surfaces of other cell types are also involved. After nerve injury, macrophages as well as other immune cell infiltrates are detected in the injury site, DRG, and spinal cord, where the injured axons terminate (Hu and McLachlan 2003). Therefore, it is possible that TLR activation in these immune cells may also contribute to chronic pain induction after nerve injury. Future studies using bone marrow chimeric mice (wild-type mice transplanted with TLR-deficient mouse bone marrow or vice versa) would address this issue (Kang and Rivest 2007). As described in the previous section, Schwann cells are activated by nerve injury and produce various proinflammatory mediators that have the potential to sensitize sensory neurons. Considering the putative role of TLRs in nerve injury-induced Schwann cell activation, it can also be conjectured that the reduction in pain-mediating gene expression in TLR-deficient Schwann cells may affect chronic pain induction after nerve injury. This possibility warrants future investigation. In addition, recent reports show that some TLR members are also expressed in cortical and sensory neurons (Cameron et al. 2007; Ma et al. 2006; Tang et al. 2007; Wadachi and Hargreaves 2006). Thus far, TLR expression in spinal cord neurons has not been documented. Still, it cannot be formally excluded that spinal cord and sensory neuronal TLRs are involved in nerve injury-induced pain hypersensitivity. Figure 1 depicts possible models for how TLRs might be involved in nerve injury and neuropathic pain.

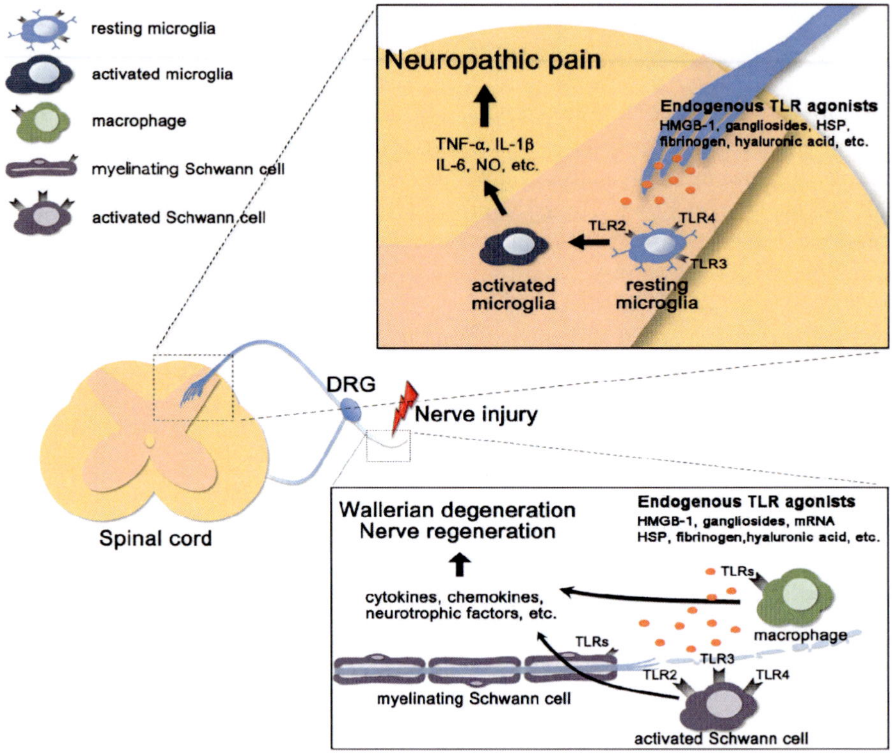

Fig. 1 TLRs in peripheral nerve injury and neuropathic pain

3.4 *Endogenous TLR Ligands Involved in Nerve Injury and Neuropathic Pain*

Thus far, the endogenous TLR ligands released after peripheral nerve injury have not been identified. Based on the literature, there are several candidate molecules that have been reported to be released from damaged neurons and to function as TLR endogenous agonists. For example, HMGB-1, a previously known DNA binding protein, is released from damaged neurons in vitro and from ischemic mouse brain in vivo, and functions as a potent microglia activator (Kim et al. 2006). It is also known that HMGB-1 activates innate immune cells via TLRs 2 and 4 (Park et al. 2006). Based on these previous reports, it is tempting to speculate that HMGB-1 released from damaged neurons activates spinal cord microglia via TLRs 2 and 4. It has also been reported that gangliosides, sialic acid-containing glycosphingolipids enriched in the neuronal membrane, function as endogenous agonists of TLRs 2 and 4 (Jou et al. 2006; Yoon et al. 2008). Therefore, it is also conceivable that peripheral nerve injury induces the release of gangliosides in the spinal cord and injured nerve, thereby activating microglia and Schwann cells. In addition, it is possible that other endogenous TLR2 or TLR4 agonists that have been previously reported to activate innate immune cells, including heat-shock proteins, fibrinogen and hyaluronic acid (Smiley et al. 2001; Termeer et al. 2002; Vabulas et al. 2002), may function as glial cell activators. mRNA molecules released from necrotic cells are known to activate TLR3 (Cavassani et al. 2008; Kariko et al. 2004). Therefore, mRNA released from injured nerves may activate nearby Schwann cells and macrophages. Likewise, mRNA, if it is released in the spinal cord after nerve injury, may activate spinal cord microglia and astrocytes. All of these speculations will be addressed in future studies in which the release of the above candidate molecules will be directly assessed in the spinal cord and injured nerve after nerve injury.

4 Summary

Schwann cells and microglia respond to nerve injury, and play key roles in the regulation of nerve degeneration and the development of neuropathic pain, respectively. Studies on the function of TLRs in these glial cells suggest that TLRs 2, 3, and 4 may function as receptors sensing nerve injury. In addition, studies using knockout mice demonstrated that TLRs 2 and 4 are required for nerve injury-induced Wallerian degeneration, and that TLRs 2, 3, and 4 are involved in the development of neuropathic pain after nerve injury. The mechanisms underlying the role of TLRs in nerve injury still need to be elucidated in future studies. In addition, given that other TLR members also recognize endogenous ligands, studies on the role of other TLRs on nerve degeneration and neuropathic pain are warranted. A large number of people worldwide are affected by nerve injury and experience disability due to unsuccessful nerve regeneration, and many of these patients suffer from chronic pain. Current management strategies for neuropathic pain that target the pain-transmitting neurons have mostly failed, and efforts are being made to develop new strategies by regulating microglia activation and subsequent inflammatory responses in the spinal cord. The critical role of TLRs in nerve injury-induced microglia activation and pain hypersensitivity suggests that these genes could be novel therapeutic targets for the treatment of neuropathic pain.

Acknowledgments This work was supported by a Korea Research Foundation Grant (KRF-2008-313-C00749) and a Korea Science and Engineering Foundation (KOSEF) grant funded by the Korean government (MOST) (R01-2007-000-10243-0, R11-2008-036-02003-0).

References

Applequist SE, Wallin RP, Ljunggren HG (2002) Variable expression of Toll-like receptor in murine innate and adaptive immune cell lines. Int Immunol 14:1065–1074

Arruda JL, Sweitzer S, Rutkowski MD, DeLeo JA (2000) Intrathecal anti-IL-6 antibody and IgG attenuates peripheral nerve injury-induced mechanical allodynia in the rat: possible immune modulation in neuropathic pain. Brain Res 879:216–225

Babcock AA, Wirenfeldt M, Holm T, Nielsen HH, Dissing-Olesen L, Toft-Hansen H, Millward JM, Landmann R, Rivest S, Finsen B, Owens T (2006) Toll-like receptor 2 signalingin response to brain injury: an innate bridge to neuroinflammation. J Neurosci 2612826–12837

Bennett GJ, Xie YK (1988) A peripheral mononeuropathy in rat that produces disorders of pain sensation like those seen in man. Pain 33:87–107

Boivin A, Pineau I, Barrette B, Filali M, Vallieres N, Rivest S, Lacroix S (2007) Toll-like receptor signaling is critical for Wallerian degeneration and functional recovery after peripheral nerve injury. J Neurosci 27:12565–12576

Bolin LM, Verity AN, Silver JE, Shooter EM, Abrams JS (1995) Interleukin-6 production by Schwann cells and induction in sciatic nerve injury. J Neurochem 64:850–858

Boyd JG, Gordon T (2003) Glial cell line-derived neurotrophic factor and brain-derived neurotrophic factor sustain the axonal regeneration of chronically axotomized motoneurons in vivo. Exp Neurol 183:610–619

Broom DC, Samad TA, Kohno T, Tegeder I, Geisslinger G, Woolf CJ (2004) Cyclooxygenase 2 expression in the spared nerve injury model of neuropathic pain. Neuroscience 124:891–900

Cameron JS, Alexopoulou L, Sloane JA, DiBernardo AB, Ma Y, Kosaras B, Flavell R, Strittmatter SM, Volpe J, Sidman R, Vartanian T (2007) Toll-like receptor 3 is a potent negative regulator of axonal growth in mammals. J Neurosci 27:13033–13041

Carroll SL, Frohnert PW (1998) Expression of JE (monocyte chemoattractant protein-1) is induced by sciatic axotomy in wild type rodents but not in C57BL/Wld(s) mice. J Neuropathol Exp Neurol 57:915–930

Cavassani KA, Ishii M, Wen H, Schaller MA, Lincoln PM, Lukacs NW, Hogaboam CM, Kunkel SL (2008) TLR3 is an endogenous sensor of tissue necrosis during acute inflammatory events. J Exp Med 205:2609–2621

Chakravarty S, Herkenham M (2005) Toll-like receptor 4 on nonhematopoietic cells sustains CNS inflammation during endotoxemia, independent of systemic cytokines. J Neurosci 25: 1788–1796

Chattopadhyay S, Myers RR, Janes J, Shubayev V (2007) Cytokine regulation of MMP-9 in peripheral glia: implications for pathological processes and pain in injured nerve. Brain Behav Immun 21:561–568

Colburn RW, Rickman AJ, DeLeo JA (1999) The effect of site and type of nerve injury on spinal glial activation and neuropathic pain behavior. Exp Neurol 157:289–304

Coleman M (2005) Axon degeneration mechanisms: commonality amid diversity. Nat Rev Neurosci 6:889–898

Coull JA, Beggs S, Boudreau D, Boivin D, Tsuda M, Inoue K, Gravel C, Salter MW, De Koninck Y (2005) BDNF from microglia causes the shift in neuronal anion gradient underlying neuropathic pain. Nature 438:1017–1021

Cui JG, Holmin S, Mathiesen T, Meyerson BA, Linderoth B (2000) Possible role of inflammatory mediators in tactile hypersensitivity in rat models of mononeuropathy. Pain 88:239–248

Cunha TM, Verri WA Jr, Silva JS, Poole S, Cunha FQ, Ferreira SH (2005) A cascade of cytokines mediates mechanical inflammatory hypernociception in mice. Proc Natl Acad Sci USA 102:1755–1760

Daffis S, Samuel MA, Suthar MS, Gale M, Diamond MS (2008) Toll-like receptor 3 has a protective role against west nile virus infection. J Virol 82:10349–10358

Decosterd I, Woolf CJ (2000) Spared nerve injury: an animal model of persistent peripheral neuropathic pain. Pain 87:149–158

DeLeo JA, Colburn RW, Nichols M, Malhotra A (1996) Interleukin-6-mediated hyperalgesia/allodynia and increased spinal IL-6 expression in a rat mononeuropathy model. J Interferon Cytokine Res 16:695–700

Dib-Hajj SD, Fjell J, Cummins TR, Zheng Z, Fried K, LaMotte R, Black JA, Waxman SG (1999) Plasticity of sodium channel expression in DRG neurons in the chronic constriction injury model of neuropathic pain. Pain 83:591–600

Dowsing BJ, Romeo R, Morrison WA (2001) Expression of leukemia inhibitory factor in human nerve following injury. J Neurotrauma 18:1279–1287

Gilmore SA, Skinner RD (1979) Intraspinal non-neuronal cellular responses to peripheral nerve injury. Anat Rec 194:369–387
Guertin AD, Zhang DP, Mak KS, Alberta JA, Kim HA (2005) Microanatomy of axon/glial signaling during Wallerian degeneration. J Neurosci 25:3478–3487
Hammarberg H, Piehl F, Cullheim S, Fjell J, Hokfelt T, Fried K (1996) GDNF mRNA in Schwann cells and DRG satellite cells after chronic sciatic nerve injury. Neuroreport 7:857–860
Hanz S, Perlson E, Willis D, Zheng JQ, Massarwa R, Huerta JJ, Koltzenburg M, Kohler M, van-Minnen J, Twiss JL, Fainzilber M (2003) Axoplasmic importins enable retrograde injury signaling in lesioned nerve. Neuron 40:1095–1104
Hashizume H, DeLeo JA, Colburn RW, Weinstein JN (2000) Spinal glial activation and cytokine expression after lumbar root injury in the rat. Spine 25:1206–1217
Hu P, McLachlan EM (2003) Distinct functional types of macrophage in dorsal root ganglia and spinal nerves proximal to sciatic and spinal nerve transections in the rat. Exp Neurol 184:590–605
Hutchinson MR, Zhang Y, Brown K, Coats BD, Shridhar M, Sholar PW, Patel SJ, Crysdale NY, Harrison JA, Maier SF, Rice KC, Watkins LR (2008) Non-stereoselective reversal of neuropathic pain by naloxone and naltrexone: involvement of toll-like receptor 4 (TLR4). Eur J Neurosci 28:20–29
Jackson AC, Rossiter JP, Lafon M (2006) Expression of Toll-like receptor 3 in the human cerebellar cortex in rabies, herpes simplex encephalitis, and other neurological diseases. J Neurovirol 12:229–234
Jou I, Lee JH, Park SY, Yoon HJ, Joe EH, Park EJ (2006) Gangliosides trigger inflammatory responses via TLR4 in brain glia. Am J Pathol 168:1619–1630
Jungnickel J, Haase K, Konitzer J, Timmer M, Grothe C (2006) Faster nerve regeneration after sciatic nerve injury in mice over-expressing basic fibroblast growth factor. J Neurobiol 66:940–948
Kang J, Rivest S (2007) MyD88-deficient bone marrow cells accelerate onset and reduce survival in a mouse model of amyotrophic lateral sclerosis. J Cell Biol 179:1219–1230
Karanth S, Yang G, Yeh J, Richardson PM (2006) Nature of signals that initiate the immune response during Wallerian degeneration of peripheral nerves. Exp Neurol 202:161–166
Kariko K, Ni H, Capodici J, Lamphier M, Weissman D (2004) mRNA is an endogenous ligand for Toll-like receptor 3. J Biol Chem 279:12542–12550
Kerschensteiner M, Schwab ME, Lichtman JW, Misgeld T (2005) In vivo imaging of axonal degeneration and regeneration in the injured spinal cord. Nat Med 11:572–577
Kim D, Kim MA, Cho IH, Kim MS, Lee S, Jo EK, Choi SY, Park K, Kim JS, Akira S, Na HS, Oh SB, Lee SJ (2007) A critical role of toll-like receptor 2 in nerve injury-induced spinal cord glial cell activation and pain hypersensitivity. J Biol Chem 282:14975–14983
Kim JB, Sig Choi J, Yu YM, Nam K, Piao CS, Kim SW, Lee MH, Han PL, Park JS, Lee JK (2006) HMGB1, a novel cytokine-like mediator linking acute neuronal death and delayed neuroinflammation in the postischemic brain. J Neurosci 26:6413–6421
Kim SH, Chung JM (1992) An experimental model for peripheral neuropathy produced by segmental spinal nerve ligation in the rat. Pain 50:355–363
Koltzenburg M (2005) Mechanisms of peripheral neuropathic pain. In: Hunt S, Koltzenburg M (eds) The neurobiology of pain (Molecular and Cellular Neurobiology Series), 1st edn. Oxford University Press, Oxford
Kuhn G, Lie A, Wilms S, Muller HW (1993) Coexpression of *PMP22* gene with MBP and P0 during de novo myelination and nerve repair. Glia 8:256–264
Lafon M, Megret F, Lafage M, Prehaud C (2006) The innate immune facet of brain: human neurons express TLR-3 and sense viral dsRNA. J Mol Neurosci 29:185–194
Ledeboer A, Sloane EM, Milligan ED, Frank MG, Mahony JH, Maier SF, Watkins LR (2005) Minocycline attenuates mechanical allodynia and proinflammatory cytokine expression in rat models of pain facilitation. Pain 115:71–83
Lee H, Jo EK, Choi SY, Oh SB, Park K, Kim JS, Lee SJ (2006) Necrotic neuronal cells induce inflammatory Schwann cell activation via TLR2 and TLR3: implication in Wallerian degeneration. Biochem Biophys Res Commun 350:742–747

Lee H, Park C, Cho IH, Kim HY, Jo EK, Lee S, Kho HS, Choi SY, Oh SB, Park K, Kim JS, Lee SJ (2007) Double-stranded RNA induces iNOS gene expression in Schwann cells, sensory neuronal death, and peripheral nerve demyelination. Glia 55:712–722

Lehnardt S, Lachance C, Patrizi S, Lefebvre S, Follett PL, Jensen FE, Rosenberg PA, Volpe JJ, Vartanian T (2002) The toll-like receptor TLR4 is necessary for lipopolysaccharide-induced oligodendrocyte injury in the CNS. J Neurosci 22:2478–2486

Levy D, Hoke A, Zochodne DW (1999) Local expression of inducible nitric oxide synthase in an animal model of neuropathic pain. Neurosci Lett 260:207–209

Levy D, Zochodne DW (1998) Local nitric oxide synthase activity in a model of neuropathic pain. Eur J Neurosci 10:1846–1855

Li M, Carpio DF, Zheng Y, Bruzzo P, Singh V, Ouaaz F, Medzhitov RM, Beg AA (2001) An essential role of the NF-kappa B/Toll-like receptor pathway in induction of inflammatory and tissue-repair gene expression by necrotic cells. J Immunol 166:7128–7135

Lindenlaub T, Teuteberg P, Hartung T, Sommer C (2000) Effects of neutralizing antibodies to TNF-alpha on pain-related behavior and nerve regeneration in mice with chronic constriction injury. Brain Res 866:15–22

Lindwall C, Kanje M (2005) Retrograde axonal transport of JNK signaling molecules influence injury induced nuclear changes in p-c-Jun and ATF3 in adult rat sensory neurons. Mol Cell Neurosci 29:269–282

Luk HW, Noble LJ, Werb Z (2003) Macrophages contribute to the maintenance of stable regenerating neurites following peripheral nerve injury. J Neurosci Res 73:644–658

Ma W, Du W, Eisenach JC (2002) Role for both spinal cord COX-1, COX-2 in maintenance of mechanical hypersensitivity following peripheral nerve injury. Brain Res 937:94–99

Ma Y, Li J, Chiu I, Wang Y, Sloane JA, Lu J, Kosaras B, Sidman RL, Volpe JJ, Vartanian T (2006) Toll-like receptor 8 functions as a negative regulator of neurite outgrowth and inducer of neuronal apoptosis. J Cell Biol 175:209–215

Mack TG, Reiner M, Beirowski B, Mi W, Emanuelli M, Wagner D, Thomson D, Gillingwater T, Court F, Conforti L, Fernando FS, Tarlton A, Andressen C, Addicks K, Magni G, Ribchester RR, Perry VH, Coleman MP (2001) Wallerian degeneration of injured axons and synapses is delayed by a *Ube4b/Nmnat* chimeric gene. Nat Neurosci 4:1199–1206

Malin SA, Molliver DC, Koerber HR, Cornuet P, Frye R, Albers KM, Davis BM (2006) Glial cell line-derived neurotrophic factor family members sensitize nociceptors in vitro and produce thermal hyperalgesia in vivo. J Neurosci 26:8588–8599

Malmberg AB, Bannon AW (1999) Models of nociception: hot-plate, tail-flick, and formalin tests in rodents. In: Taylor G (ed) Current protocols in neuroscience. Wiley, Hoboken

Marchand F, Perretti M, McMahon SB (2005) Role of the immune system in chronic pain. Nat Rev Neurosci 6:521–532

Matzinger P (2002) The danger model: a renewed sense of self. Science 296:301–305

McMahon SB, Cafferty WB, Marchand F (2005) Immune and glial cell factors as pain mediators and modulators. Exp Neurol 192:444–462

Medzhitov R (2001) Toll-like receptors and innate immunity. Nat Rev Immunol 1:135–145

Meller ST, Pechman PS, Gebhart GF, Maves TJ (1992) Nitric oxide mediates the thermal hyperalgesia produced in a model of neuropathic pain in the rat. Neuroscience 50:7–10

Merskey H, Bogduk M (1994) Classification of chronic pain. IASP, Seattle

Meyer M, Matsuoka I, Wetmore C, Olson L, Thoenen H (1992) Enhanced synthesis of brain-derived neurotrophic factor in the lesioned peripheral nerve: different mechanisms are responsible for the regulation of BDNF and NGF mRNA. J Cell Biol 119:45–54

Milligan ED, O'Connor KA, Nguyen KT, Armstrong CB, Twining C, Gaykema RP, Holguin A, Martin D, Maier SF, Watkins LR (2001) Intrathecal HIV-1 envelope glycoprotein gp120 induces enhanced pain states mediated by spinal cord proinflammatory cytokines. J Neurosci 21:2808–2819

Mukhopadhyay G, Doherty P, Walsh FS, Crocker PR, Filbin MT (1994) A novel role for myelin-associated glycoprotein as an inhibitor of axonal regeneration. Neuron 13:757–767

Obata K, Katsura H, Miyoshi K, Kondo T, Yamanaka H, Kobayashi K, Dai Y, Fukuoka T, Akira S, Noguchi K (2008) Toll-like receptor 3 contributes to spinal glial activation and tactile allodynia after nerve injury. J Neurochem 105:2249–2259

Ohtori S, Takahashi K, Moriya H, Myers RR (2004) TNF-alpha and TNF-alpha receptor type 1 upregulation in glia and neurons after peripheral nerve injury: studies in murine DRG and spinal cord. Spine 29:1082–1088

Oliveira RB, Ochoa MT, Sieling PA, Rea TH, Rambukkana A, Sarno EN, Modlin RL (2003) Expression of Toll-like receptor 2 on human Schwann cells: a mechanism of nerve damage in leprosy. Infect Immun 71:1427–1433

Pardo CA, McArthur JC, Griffin JW (2001) HIV neuropathy: insights in the pathology of HIV peripheral nerve disease. J Peripher Nerv Syst 6:21–27

Park JS, Gamboni-Robertson F, He Q, Svetkauskaite D, Kim JY, Strassheim D, Sohn JW, Yamada S, Maruyama I, Banerjee A, Ishizaka A, Abraham E (2006) High mobility group box 1 protein interacts with multiple Toll-like receptors. Am J Physiol Cell Physiol 290:C917–C924

Poltorak A, He X, Smirnova I, Liu MY, Van Huffel C, Du X, Birdwell D, Alejos E, Silva M, Galanos C, Freudenberg M, Ricciardi-Castagnoli P, Layton B, Beutler B (1998) Defective LPS signaling in C3H/HeJ and C57BL/10ScCr mice: mutations in *Tlr4* gene. Science 282: 2085–2088

Prehaud C, Megret F, Lafage M, Lafon M (2005) Virus infection switches TLR-3-positive human neurons to become strong producers of beta interferon. J Virol 79:12893–12904

Raghavendra V, Tanga F, DeLeo JA (2003) Inhibition of microglial activation attenuates the development but not existing hypersensitivity in a rat model of neuropathy. J Pharmacol Exp Ther 306:624–630

Ramer MS, Bisby MA (1997) Rapid sprouting of sympathetic axons in dorsal root ganglia of rats with a chronic constriction injury. Pain 70:237–244

Rind HB, Butowt R, von Bartheld CS (2005) Synaptic targeting of retrogradely transported trophic factors in motoneurons: comparison of glial cell line-derived neurotrophic factor, brain-derived neurotrophic factor, and cardiotrophin-1 with tetanus toxin. J Neurosci 25: 539–549

Salter MW, Woolf CJ (2005) Cellular and molecular mechanisms of central sensitization. In: Hunt S, Koltzenburg M (eds) The neurobiology of pain (Molecular and Cellular Neurobiology Series), 1st edn. Oxford University Press, Oxford

Schweizer U, Gunnersen J, Karch C, Wiese S, Holtmann B, Takeda K, Akira S, Sendtner M (2002) Conditional gene ablation of Stat3 reveals differential signaling requirements for survival of motoneurons during development and after nerve injury in the adult. J Cell Biol 156:287–297

Seltzer Z, Dubner R, Shir Y (1990) A novel behavioral model of neuropathic pain disorders produced in rats by partial sciatic nerve injury. Pain 43:205–218

Shubayev VI, Angert M, Dolkas J, Campana WM, Palenscar K, Myers RR (2006) TNFalpha-induced MMP-9 promotes macrophage recruitment into injured peripheral nerve. Mol Cell Neurosci 31:407–415

Skundric DS, Bealmear B, Lisak RP (1997) Induced upregulation of IL-1, IL-1RA and IL-1R type I gene expression by Schwann cells. J Neuroimmunol 74:9–18

Smiley ST, King JA, Hancock WW (2001) Fibrinogen stimulates macrophage chemokine secretion through toll-like receptor 4. J Immunol 167:2887–2894

Smith GM, Rabinovsky ED, McManaman JL, Shine HD (1993) Temporal and spatial expression of ciliary neurotrophic factor after peripheral nerve injury. Exp Neurol 121:239–247

Sommer C, Lindenlaub T, Teuteberg P, Schafers M, Hartung T, Toyka KV (2001) Anti-TNF-neutralizing antibodies reduce pain-related behavior in two different mouse models of painful mononeuropathy. Brain Res 913:86–89

Stoll G, Jander S, Myers RR (2002) Degeneration and regeneration of the peripheral nervous system: from Augustus Waller's observations to neuroinflammation. J Peripher Nerv Syst 7:13–27

Stoll G, Jung S, Jander S, van der Meide P, Hartung HP (1993) Tumor necrosis factor-alpha in immune-mediated demyelination and Wallerian degeneration of the rat peripheral nervous system. J Neuroimmunol 45:175–182

Sugimoto T, Bennett GJ, Kajander KC (1990) Transsynaptic degeneration in the superficial dorsal horn after sciatic nerve injury: effects of a chronic constriction injury, transection, and strychnine. Pain 42:205–213

Sweitzer S, Martin D, DeLeo JA (2001) Intrathecal interleukin-1 receptor antagonist in combination with soluble tumor necrosis factor receptor exhibits an anti-allodynic action in a rat model of neuropathic pain. Neuroscience 103:529–539

Takahashi M, Kawaguchi M, Shimada K, Konishi N, Furuya H, Nakashima T (2004) Cyclooxygenase-2 expression in Schwann cells and macrophages in the sciatic nerve after single spinal nerve injury in rats. Neurosci Lett 363:203–206

Tang SC, Arumugam TV, Xu X, Cheng A, Mughal MR, Jo DG, Lathia JD, Siler DA, Chigurupati S, Ouyang X, Magnus T, Camandola S, Mattson MP (2007) Pivotal role for neuronal Toll-like receptors in ischemic brain injury and functional deficits. Proc Natl Acad Sci USA 104: 13798–13803

Tanga FY, Nutile-McMenemy N, DeLeo JA (2005) The CNS role of Toll-like receptor 4 in innate neuroimmunity and painful neuropathy. Proc Natl Acad Sci USA 102:5856–5861

Tanga FY, Raghavendra V, DeLeo JA (2004) Quantitative real-time RT-PCR assessment of spinal microglial and astrocytic activation markers in a rat model of neuropathic pain. Neurochem Int 45:397–407

Taskinen HS, Roytta M (2000) Increased expression of chemokines (MCP-1, MIP-1alpha, RANTES) after peripheral nerve transection. J Peripher Nerv Syst 5:75–81

Termeer C, Benedix F, Sleeman J, Fieber C, Voith U, Ahrens T, Miyake K, Freudenberg M, Galanos C, Simon JC (2002) Oligosaccharides of hyaluronan activate dendritic cells via toll-like receptor 4. J Exp Med 195:99–111

Tofaris GK, Patterson PH, Jessen KR, Mirsky R (2002) Denervated Schwann cells attract macrophages by secretion of leukemia inhibitory factor (LIF) and monocyte chemoattractant protein-1 in a process regulated by interleukin-6 and LIF. J Neurosci 22:6696–6703

Town T, Jeng D, Alexopoulou L, Tan J, Flavell RA (2006) Microglia recognize double-stranded RNA via TLR3. J Immunol 176:3804–3812

Tsuda M, Inoue K, Salter MW (2005) Neuropathic pain and spinal microglia: a big problem from molecules in "small" glia. Trends Neurosci 28:101–107

Vabulas RM, Wagner H, Schild H (2002) Heat shock proteins as ligands of toll-like receptors. Curr Top Microbiol Immunol 270:169–184

Vikman KS, Duggan AW, Siddall PJ (2007) Interferon-gamma induced disruption of GABAergic inhibition in the spinal dorsal horn in vivo. Pain 133:18–28

Wadachi R, Hargreaves KM (2006) Trigeminal nociceptors express TLR-4 and CD14: a mechanism for pain due to infection. J Dent Res 85:49–53

Wagner R, Myers RR (1996) Schwann cells produce tumor necrosis factor alpha: expression in injured and non-injured nerves. Neuroscience 73:625–629

Watkins LR, Milligan ED, Maier SF (2001) Glial activation: a driving force for pathological pain. Trends Neurosci 24:450–455

Yang EK, Takimoto K, Hayashi Y, de Groat WC, Yoshimura N (2004) Altered expression of potassium channel subunit mRNA and alpha-dendrotoxin sensitivity of potassium currents in rat dorsal root ganglion neurons after axotomy. Neuroscience 123:867–874

Yoon HJ, Jeon SB, Suk K, Choi DK, Hong YJ, Park EJ (2008) Contribution of TLR2 to the initiation of ganglioside-triggered inflammatory signaling. Mol Cells 25:99–104

Yura Y (2000) Orofacial alpha herpesvirus infection. Nippon Rinsho 58:912–917

Zochodne DW, Levy D, Zwiers H, Sun H, Rubin I, Cheng C, Lauritzen M (1999) Evidence for nitric oxide and nitric oxide synthase activity in proximal stumps of transected peripheral nerves. Neuroscience 91:1515–1527

Index

M

N

O

P